Studies in Computational Intelligence

Volume 1052

Series Editor
Janusz Kacprzyk, Polish Academy of Sciences, Warsaw, Poland

The series "Studies in Computational Intelligence" (SCI) publishes new developments and advances in the various areas of computational intelligence—quickly and with a high quality. The intent is to cover the theory, applications, and design methods of computational intelligence, as embedded in the fields of engineering, computer science, physics and life sciences, as well as the methodologies behind them. The series contains monographs, lecture notes and edited volumes in computational intelligence spanning the areas of neural networks, connectionist systems, genetic algorithms, evolutionary computation, artificial intelligence, cellular automata, self-organizing systems, soft computing, fuzzy systems, and hybrid intelligent systems. Of particular value to both the contributors and the readership are the short publication timeframe and the world-wide distribution, which enable both wide and rapid dissemination of research output.

Indexed by SCOPUS, DBLP, WTI Frankfurt eG, zbMATH, SCImago.

All books published in the series are submitted for consideration in Web of Science.

Pranesh Santikellur · Rajat Subhra Chakraborty

Deep Learning for Computational Problems in Hardware Security

Modeling Attacks on Strong Physically Unclonable Function Circuits

Pranesh Santikellur
Computer Science and Engineering
Indian Institute of Technology Kharagpur
Kharagpur, West Bengal, India

Rajat Subhra Chakraborty
Computer Science and Engineering
Indian Institute of Technology Kharagpur
Kharagpur, West Bengal, India

ISSN 1860-949X ISSN 1860-9503 (electronic)
Studies in Computational Intelligence
ISBN 978-981-19-4016-3 ISBN 978-981-19-4017-0 (eBook)
https://doi.org/10.1007/978-981-19-4017-0

© The Editor(s) (if applicable) and The Author(s), under exclusive license to Springer Nature Singapore Pte Ltd. 2023
This work is subject to copyright. All rights are solely and exclusively licensed by the Publisher, whether the whole or part of the material is concerned, specifically the rights of translation, reprinting, reuse of illustrations, recitation, broadcasting, reproduction on microfilms or in any other physical way, and transmission or information storage and retrieval, electronic adaptation, computer software, or by similar or dissimilar methodology now known or hereafter developed.
The use of general descriptive names, registered names, trademarks, service marks, etc. in this publication does not imply, even in the absence of a specific statement, that such names are exempt from the relevant protective laws and regulations and therefore free for general use.
The publisher, the authors, and the editors are safe to assume that the advice and information in this book are believed to be true and accurate at the date of publication. Neither the publisher nor the authors or the editors give a warranty, expressed or implied, with respect to the material contained herein or for any errors or omissions that may have been made. The publisher remains neutral with regard to jurisdictional claims in published maps and institutional affiliations.

This Springer imprint is published by the registered company Springer Nature Singapore Pte Ltd.
The registered company address is: 152 Beach Road, #21-01/04 Gateway East, Singapore 189721, Singapore

Dedicated to,
Almighty "Gandhavahana", Appa, Amma,
my wife Anusha, our daughter Apeksha and
family members.

— Pranesh Santikellur

Dedicated to,
My wife Munmun and our daughter Agomoni.

— Dr. Rajat Subhra Chakraborty

Preface

Recent years have witnessed an increased focus on hardware security research. There have been significant advances in hardware security over the last decade. A recent discovery of chip vulnerabilities such as *Spectre* and *Meltdown* shows that the foundation of trust in hardware can also be insecure. Additionally, the tremendous growth of machine learning can either be a boon or a bane for hardware security. In the field of hardware security, Physically Unclonable Functions (PUFs) are being considered as a potential root of trust for solving hardware security problems. However, machine learning-based modeling attacks have been the greatest threat to PUF.

This book describes machine-learning applications for hardware security, focusing on modeling attacks against PUFs. The contents of this book are the results of our intensive research on the above-mentioned topic over the last few years. This book presents a practical guide for the use of machine learning in modeling attacks on PUF. We wrote this book to share our experience of research on modeling attacks on PUFs in a systematic and comprehensive way that covers both depth and breadth of the topic. Since this topic has not been covered in a single volume till date, we feel it would be of great value to any interested reader. The topics have been covered in sufficient depth to enable the reader to develop the understanding and skill-sets to explore it further. We have released the source-code corresponding to the PUF models and the modeling attacks in the public domain, to further aid the readers who would be interested in exploring this fascinating topic and enhancing the state of the art through future research.

Whether you are a complete beginner or an experienced researcher, the book brings together a formidable amount of theory and practical know-how under a single cover and will assist you in recollecting and gaining the knowledge of modeling attacks in the most comprehensive way.

Kharagpur, India
September 2021

Pranesh Santikellur
Dr. Rajat Subhra Chakraborty

Acknowledgements Pranesh Santikellur would like to acknowledge the financial support provided to him by Intel.

Contents

1 **Introduction to Machine Learning for Hardware Security** 1
   1.1 Machine Learning: How is it Different from Traditional Algorithms? 3
   1.2 Machine Learning for Hardware Security 4
      1.2.1 ML for Hardware Attacks 4
      1.2.2 ML for Hardware Attack Detection 5
      1.2.3 ML for Attack on Countermeasures 6
   1.3 Organization of the Book 6
   References 7

2 **Physically Unclonable Functions** 9
   2.1 Introduction 9
      2.1.1 Properties of PUF 10
      2.1.2 Quality Metrics of PUF 11
      2.1.3 PUF Classification 12
   2.2 Arbiter PUF (APUF) 13
      2.2.1 Delay Model of Arbiter PUF 14
   2.3 PUF Composition Types 14
      2.3.1 XOR PUF Composition 16
      2.3.2 Tribes PUF Composition 18
      2.3.3 Multiplexer PUF (MPUF) and Its Variants 19
      2.3.4 Interpose PUF (IPUF) 20
   References 21

3 **Machine-Learning Basics** 23
   3.1 Introduction 23
   3.2 Machine-Learning Classification 24
   3.3 Supervised ML Algorithms 25
      3.3.1 Support Vector Machines 25
      3.3.2 Logistic Regression 28
      3.3.3 Artificial Neural Networks 30

	3.4	Unsupervised ML Algorithms	33
		3.4.1 K-means Clustering	33
	References		34

4 Modeling Attacks on PUF ... 35
- 4.1 Introduction ... 35
- 4.2 Mathematical Model of Arbiter PUF (APUF) ... 36
- 4.3 Mathematical Model of XOR APUF ... 38
    - 4.3.1 Motivation of Applying Deep Learning to PUF Modeling Attack ... 39
- 4.4 DFNN Architecture for Modeling Attack on APUF Compositions ... 40
    - 4.4.1 Modeling Attack on XOR APUF ... 42
    - 4.4.2 Modeling Attack on Double Arbiter PUF (DAPUF) ... 45
    - 4.4.3 Modeling Attack on Multiplexer PUF and Its Variants ... 46
    - 4.4.4 Modeling Attack on Interpose PUF (iPUF) ... 46
- References ... 52

5 Improved Modeling Attack on PUFs based on Tensor Regression Network ... 55
- 5.1 Introduction ... 55
- 5.2 Tensor Basics ... 56
    - 5.2.1 CP-Decomposition ... 57
    - 5.2.2 Tensor Regression Networks ... 58
- 5.3 ECP-TRN: Efficient CP-Decomposition-Based Tensor Regression Networks ... 58
- 5.4 Experimental Results ... 61
    - 5.4.1 Simulation and Modeling Setup ... 61
    - 5.4.2 Modeling Accuracy Results ... 62
- 5.5 ECP-TRN-Based Modeling Attack XOR APUF Variants ... 65
    - 5.5.1 ECP-TRN-Based Modeling Attack on Lightweight Secure PUF (LSPUF) ... 66
    - 5.5.2 ECP-TRN-Based Modeling Attack on PC-XOR APUF ... 67
    - 5.5.3 ECP-TRN-Based Modeling Attack on Mixed Challenge XOR APUF ... 67
- References ... 68

6 Combinational Logic-Based Implementation of PUF ... 71
- 6.1 Introduction ... 71
    - 6.1.1 ML-Based PUF Models in PUF-Based Authentication Protocols ... 72
- 6.2 Binarized Neural Networks ... 73
- 6.3 Optimized Combinational Logic-Based BNN Implementation ... 75

		6.4	The APUF-BNN CAD Framework	77

- 6.4 The APUF-BNN CAD Framework 77
  - 6.4.1 BNN-Based Modeling Attack on APUF 78
  - 6.4.2 Matrix Covering-Based Logic Optimization and Combinational Verilog Code Generation 78
- 6.5 Experimental Results 79
  - 6.5.1 Setup 79
  - 6.5.2 Experimental Results 79
- References 81

7 Conclusion 83

About the Authors

Pranesh Santikellur is a Ph.D. student and a Senior Research Fellow in the Department of Computer Science and Engineering at the Indian Institute of Technology, Kharagpur. He received his B.E. degree in Electronics & Communication Engineering from Visvesvaraya Technological University, Belgaum, India, in 2010. He has a total of 6 years of industry experience at Horner Engineering India Pvt. Ltd. and Processor Systems. His primary research interest lies in hardware security, deep learning, and programmable logic controller security. He is an IEEE student member.

Rajat Subhra Chakraborty is an Associate Professor in the Department of Computer Science & Engineering of the Indian Institute of Technology, Kharagpur, India. He has professional experience working in National Semiconductor and Advanced Micro Devices (AMD). His research interest lies in the areas of hardware security, VLSI design, digital watermarking, and digital image forensics, in which he has published 4 books and over 100 papers in international journals and conferences of repute. He holds 2 granted U.S. patents. His publications have received over 3600 citations to date. Dr. Chakraborty has a Ph.D. in Computer Engineering from Case Western Reserve University, USA, and is a senior member of IEEE and ACM.

Chapter 1
Introduction to Machine Learning for Hardware Security

Computing devices have become the basis of our modern lives, and hardware has long been considered as the backbone of trust for all computing systems. Various software attacks and defense mechanisms, mainly based on cryptographic measures, have been widely analyzed and applied in a variety of applications. In comparison to software security, hardware security as a topic is relatively new, and its importance has drastically increased in recent years due to the multiple attacks on hardware that were thought to be immune to attack. Hardware security is no different than any other field of security that focuses on launching attacks to steal assets and on strategies designed to protect them. In particular, the topic of hardware security is focused on situations where the assets are hardware components that contain secrets of electronic components, such as cryptographic keys and other sensitive information [1].

Early efforts in hardware security were mainly aimed at exploring and defending against attacks which leak sensitive information by finding implementation-based vulnerabilities in cryptographic hardware. However, it has become a more significant security concern recently because IP-based SoC (intellectual-property-based system on chip) designs are increasingly prevalent and fewer controls on the design and fabrication of electronic components are facilitating it. This reduced control over design and fabrication can be attributed to the globalized nature of the supply chain for electronic components which makes it more likely to have a malicious modification of ICs in an untrusted design house or foundry; such malicious designs are called *hardware Trojans* [1]. Hardware Trojans can pose the greatest threat by allowing adversaries to alter their IC functionality and causing significant damage.

The side-channel attack is another important hardware security issue that allows the secret information stored on chips to be extracted by analyzing physical signals such as propagation delays, power levels, and electromagnetic emissions [1]. It has been demonstrated that side-channel techniques are capable of detecting the secret key and cracking robust cryptographic algorithms and therefore pose a serious threat. The other important hardware-based threats are IP piracy, reverse engineering, and

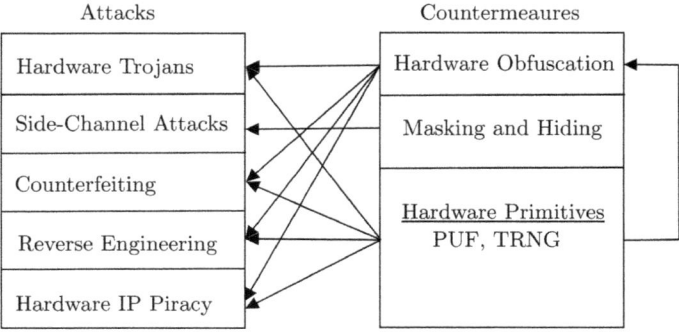

Fig. 1.1 Hardware attacks and countermeasures

counterfeiting of IC. Furthermore, hardware cloning is one of the most common attacks on PCBs, since they can be easily counterfeited by reverse engineering. Reverse engineering becomes more severe threat when its possibility is considered during every stage of IC fabrication and manufacturing.

In order to provide a defense against these emerging hardware attacks, a number of countermeasures have been proposed. *Hardware obfuscation* is one of the prominent countermeasures to combat not only IP piracy but also reverse engineering and Trojan insertion. *Logic locking* is the well-researched hardware obfuscation method that hides not only the true functionality but also structure of the hardware circuit. To prevent side-channel-based attacks, "masking and hiding" techniques has been proposed as a countermeasure that aims to reduce leakage.

In addition to hardware security research on new attacks and countermeasures, recent research has placed significant emphasis on developing trustworthy hardware for developing the root of trust. The notion of hardware as a root of trust for security is often realized through hardware building blocks, often referred to as *hardware security primitives* [1]. The new development would address trust, integrity, and authentication of devices, which could solve a number of security issues not only in hardware security but also extend its application to software security. In the development of hardware security primitives, one of the ideas is to use intrinsic properties of the hardware, like semiconductor-based process variations. These intrinsic properties, often realized through circuits, vary from device to device, even if same process is used on the same material. Due to the fact that even the same manufacturer cannot produce or clone two or more devices with the same intrinsic properties, these are called *Physically Unclonable Function* (PUF). As a result of their properties, PUFs are primarily used for key generation and device authentication. Another type of common hardware security primitive is *True Random Number Generators* (TRNGs), which are able to generate digital bitstreams with a high degree of uncertainty. In other words, the sequence of 1s and 0s they generate are not influenced by previous values. TRNGs can be used to generate random seeds, session keys, one-time pads, nonces, and challenges to PUFs, which are crucial for security and cryptographic applications. Figure 1.1 shows the different hardware platform-specific attacks, along with their countermeasures, indicated with arrows pointing toward each attack.

Over the last decade, machine learning has made significant progress in various fields, such as computer vision, bio-informatics, and machine translation. In fact, recent progress has made machine-learning algorithms to surpass human performance at various tasks such as language understanding [6] and strategic games [5]. The emergence of such breakthroughs is mainly credited to the use of *Deep learning* and *Reinforcement learning*. The use of machine learning is gaining traction in the field of hardware security and recent improvements in hardware security show it is also quite applicable and impactful in this field. Some ML applications have improved existing attacks and countermeasures, while other set of applications has come up with new attacks and countermeasures. It is exciting that machine learning improves the battle between attacks and countermeasures strategies in hardware security. In the next section, we examine machine learning briefly in order to understand its relevance to hardware security.

1.1 Machine Learning: How is it Different from Traditional Algorithms?

Computer algorithms are typically used to solve real-world problems such as sorting numbers. However, for certain tasks, we don't have an algorithm such as detection of spam email, since the definition of spam varies from person to person and also over time. There is no precise knowledge of how to detect an email document as spam or not. However, we can compensate for the lack of precise knowledge with extensive datasets, e.g., a collection of a few thousand example emails, some of which we know are spam, and we would like to "learn" from them what constitutes spam from the data. Specifically, we want the computer to identify the logic ("algorithm") behind spam detection.

In many instances, we have such a multitude of tasks and we do believe there to be a process that can explain the data we observe, although we are unaware of the details of the process. It is useful if we can construct an approximation that does not explain every aspect of the process, but can identify certain patterns or regularities, then it can be used to make predictions. Devising such approximations from the data is the essence of machine learning. Formally, "machine learning" (ML) term was coined by *Arthur Samuel* in 1959 and defined it as "field of study that gives computers the capability to learn without being explicitly programmed". Machine learning uses the theory of statistics to construct approximate mathematical models based on example data, called "training data" and the process of learning is called training. The constructed model is then used to derive *inference* from the data, be it for output prediction for unknown data or to gain insight into the data. In the next section, we will discuss potential applications of machine learning in the area of hardware security.

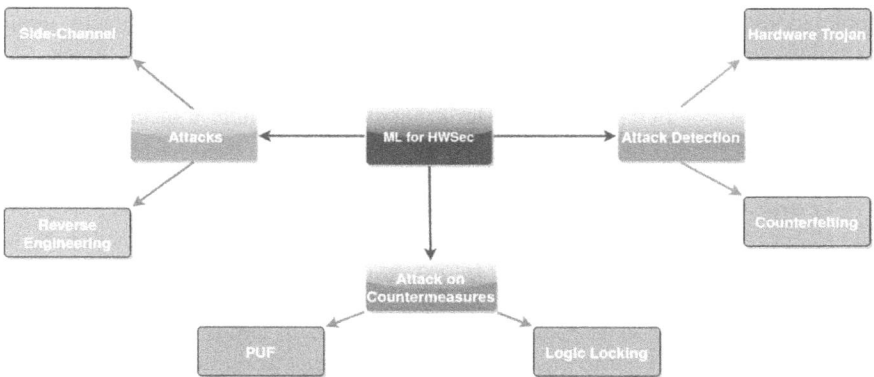

Fig. 1.2 ML for hardware security

1.2 Machine Learning for Hardware Security

We classify ML applications in hardware security into three categories: ML-assisted hardware attacks, ML-based attack detection, and ML-based attack on countermeasures. Figure 1.2 illustrates different ML applications for hardware security. In Sect. 1.1, we discussed attacks and countermeasures related to hardware security. Here, we consider "attack detection" as a different category as it does not belong to countermeasures. Next, we will discuss each category and problem ML attempts to solve.

1.2.1 ML for Hardware Attacks

We will discuss how ML is applicable side-channel attacks and reverse-engineering. In side-channel analysis, attackers analyze signals, such as radiation and power consumption, to obtain the secret keys of cryptographic algorithms. side-channel analysis can be classified mainly into two types [8]: profiling analysis and non-profiling analysis. In profiled analysis, attacker may priorly use an available copy of the target device to precisely tune all the parameters of the attack. So attacker, during training phase, can apply ML algorithms from different available side-channel traces of available target device to construct approximate mathematical model that extracts the secret key. During inference or attack phase, the attacker can launch the prediction on the trained algorithm to derived the secret keys. In non-profiling analysis, attacker does not have access to the open copy of target device and has only access to leakage of target device. Hence, they can use ML algorithms that extract directly from hidden properties of collected information. There has been also ML-based application on side-channel analysis for instruction-level disassembly. Recently we have also seen ML-based side-channel analysis on post-quantum hardware [3].

Recently, ML-assisted reverse engineering has attracted substantial research attention [2, 4]. Reverse engineering of a chip is mainly done in five stages: decapsulation, delayering, imaging, annotation, and schematic creation. The process for the last two stages, annotation, and schematic creation is usually time-consuming and complex using traditional algorithms. Therefore, ML algorithms are typically applied on features extracted from the layout images and used in various applications. Readers are suggested to refer to the work [4] for more detailed image processing and ML-assisted reverse engineering.

1.2.2 ML for Hardware Attack Detection

Machine-learning techniques have also made a great progress in the detection of common hardware security vulnerabilities, namely, hardware Trojans and counterfeit of ICs [2]. A hardware Trojan (HT) attack involves an intentional malicious modification of a circuit design such that it shows undesired circuit functionality upon deployment. A hardware Trojan detection is typically performed under two scenarios based on whether golden chips (Trojan-free chips) are available or not. When a golden chip is present, machine-learning systems can detect Trojans at different levels of abstraction, such as gate-level netlists, RTL based, and at the chip-level. An established procedure for these ML-based methods is to take advantage of the golden chip abstraction level to extract their features, and then learn from those features to detect whether or not a design contains hardware Trojans. For chip-based Trojan detection, reverse engineering is the first step, followed by machine learning applied to features extracted from layout images for Trojan detection. In the absence of golden chip, machine-learning algorithms must learn from existing designs. In such cases, a popular method followed is to transform the gate-level netlist into graph-based representations. From there, machine-learning techniques are applied to discover the outliers. To achieve this, graph neural networks are used.

When golden chip is not available, since trojan free example design is not available, machine-learning algorithm has to learn from existing design. One method followed in such case is to transform the gate-level netlist into graph-based representation, the idea is to apply ML techniques to find the outlier in the representation. In order to accomplish this task, graph specialized ML algorithms like graph neural networks or dendrogram-based clustering algorithms are used.

Counterfeiting is a new, major threat in which counterfeit or recycled ICs are sold as originals or new. Counterfeit ICs are of high concern because of the threats of system reliability degradation and US Air force has reported that an estimated 1 million counterfeits have been reported in 2012 alone [1]. One of the easiest ways to detect counterfeit ICs is visual inspection of the wear patterns, but approach is time consuming and has relatively large probability of error. Therefore, several image processing techniques, in conjunction with machine-learning techniques, can help uncover hidden common characteristics of counterfeit ICs which can aid the automatic detection process [10].

1.2.3 ML for Attack on Countermeasures

Attacks on countermeasures methods are as bad as attacks on hardware directly. They undermine the effectiveness of proposed countermeasures and indicate a redesign is necessary. The most common and significant machine learning-based attacks on countermeasures methods are Logic Locking and PUF. There have been significant research advances in this emerging field. Logic locking is a well-known method of obfuscating hardware where a set of logic gates are added to existing IP, acting as "key gates", that protect the functionality and structure of the hardware IP. The logic gates added can be combinational or sequential, and most of the existing ML-based attacks target combinatorial logic gates. Machine learning-based attacks on combinatorial logic locking can be classified into two categories [2, 9]: oracle-guided attacks and oracle-less attacks. Here, oracle-less attacks mean that an activated design is not available whereas oracle-guided means oracle-guided scenario assumes the availability of an activated IC. As a result of the absence of an activated IC in the oracle-less model, structural leakage is exploited whereas oracle-guided attacks focus on functional aspects. An oracle-guided attack involves training the ML algorithm on a sequence of I/O observations collected from an activated IC, enabling it to learn the circuit's Boolean function. Keys are then derived using the learned Boolean function. In contrast, Oracle-less ML-based attacks exploit scheme-related structural residue to identify the locking circuitry or correct key-bit value. In these types of attacks, ML algorithms are used to identify the logic structures within a post-resynthesis target netlist that correspond to locking circuitry.

PUFs are promising hardware security primitives that can redefine the way security-related applications are implemented. Their relatively simple architecture makes them a preferred choice for resource-constrained devices. Machine-learning attacks on PUF are typically known as "model-building attacks" because they construct the model of PUF which ideally should be impossible. This book is primarily about the model-building attacks on PUF, which currently constitute the greatest threat to PUF [7]. Next, we describe the organization of the book.

1.3 Organization of the Book

This book is organized into seven chapters. The content of Chaps. 2–7 is as follows:

1. Chapter 2 discusses the fundamentals of PUF and introduces various arbiter PUF compositions.
2. Chapter 3 covers the basics of machine learning and discusses several required ML algorithms typically used for model-building attacks on PUF.
3. Chapter 4 discusses modeling attacks against PUF. This chapter discusses state-of-the-art modeling attacks as well as their advantages and disadvantages.
4. Chapter 5 presents an improved modeling attack on PUF using tensor regression. The chapter discusses the design of a new tailored ML model for launching attack on a class of PUF variants called, XOR APUF variants, and the model is termed as "Efficient CP-decomposition-based Tensor Regression Network" (ECP-TRN).
5. Chapter 6 discusses the constructive aspect of modeling attacks, namely combinatorial logic-based PUF implementation. This chapter presents an end-to-end framework, *APUF-BNN*, for representing the given PUF instance into a synthesizable gate-level Verilog description.
6. Chapter 7 presents the conclusion to the book.

We hope you find the book interesting as well as enjoyable. Stay tuned for more!

References

1. Bhunia, S., & Tehranipoor, M. (2018). *Hardware security: A hands-on learning approach*. Morgan Kaufmann.
2. Elnaggar, R., & Chakrabarty, K. (2018). Machine learning for hardware security: Opportunities and risks. *Journal of Electronic Testing, 34*(2), 183–201.
3. Regazzoni, F. (2020). Machine learning and hardware security: Challenges and opportunities-invited talk. In *IEEE/ACM International Conference on Computer Aided Design (ICCAD). IEEE* (pp. 1–6).
4. Botero, U. J., et al. (2021). Hardware trust and assurance through reverse engineering: A tutorial and outlook from image analysis and machine learning perspectives. *ACM Journal on Emerging Technologies in Computing Systems (JETC), 17*(4), 1–53.
5. Vinyals, O., et al. (2019). Grandmaster level in starcraft II using multi-agent reinforcement learning. *Nature, 575*(7782), 350–354.
6. Wang, A., Pruksachatkun, Y., Nangia, N., Singh, A., Michael, J., Hill, F., Levy, O., & Bowman, S. R. (2019). *Superglue: A Stickier Benchmark for General-Purpose Language Understanding Systems*, arXiv:1905.00537.
7. Santikellur, P., Bhattacharyay, A., & Chakraborty, R. S. (2019). Deep learning based model building attacks on arbiter PUF compositions. In *Cryptology ePrint Archive* Report 2019/566, https://eprint.iacr.org/2019/566.
8. Benadjila, R., Prouff, E., Strullu, R., Cagli, E., and Dumas, C. (2018). Study of deep learning techniques for side-channel analysis and introduction to ASCAD database. In *ANSSI, France & CEA, LETI, MINATEC Campus, France*. Online verfügbar unter https://eprint.iacr.org/2018/053.pdf, zuletzt geprüft am (Vol. 22).
9. Sisejkovic, D., Reimann, L. M., Moussavi, E., Merchant, F., & Leupers, R. (2021). Logic locking at the frontiers of machine learning: A survey on developments and opportunities. arXiv:2107.01915.
10. Asadizanjani, N., Tehranipoor, M., & Forte, D. (2017). Counterfeit electronics detection using image processing and machine learning. *Journal of Physics: Conference Series, 787*(1), 012023. IOP Publishing.

Chapter 2
Physically Unclonable Functions

2.1 Introduction

Physically Unclonable Functions (PUFs) are promising hardware security primitives which can be useful in implementing lightweight authentication protocols without the need for explicit key storage. PUFs are electronic circuits that manifest the impact of nanoscale process variation induced randomness of modern semiconductor manufacturing technology [1, 2]. Each individual PUF instance should have unique characteristics, which should clearly distinguish it from other instances of the same PUF family. Typically, this unique characteristic is in the form of the truth-table for a given PUF instance, and an n-bit input, m-bit output PUF instance can be mathematically represented as a Boolean function $f : \{0, 1\}^n \rightarrow \{0, 1\}^m$.

In PUF terminology, the input stimulus and corresponding outputs are termed "challenge" and "response", respectively. Figure 2.1 shows the PUF as a black-box Boolean mapping of challenges to responses. An n-bit input to a PUF instance, and the corresponding m-bit output generated by the PUF instance, together constitute a "Challenge-Response Pair" (CRP). A set of CRPs form a Boolean function truth-table, which uniquely identifies the corresponding PUF instance. Note that entries of this truth-table can only be inferred *after* the PUF circuit instance has been fabricated in silicon and characterized. Given the relatively large number of PUF inputs (usually ≥ 64), it is not practically feasible to exhaustively characterize a PUF instance. Hence, usually the truth-table (and thus the corresponding PUF Boolean function), is *incompletely-specified*. Next, we discuss the several important properties of the PUFs.

Fig. 2.1 PUF as a black-box boolean mapping function

2.1.1 Properties of PUF

The properties of a PUF are mainly defined in terms of its challenge-response behavior. Let us denote the challenges and responses of the PUF instance i, PUF_i as **C** and response as R.

(1) Evaluatable: To be practically usable, a PUF instance should be *evaluatable*, i.e., it should be reasonably easy to apply a challenge $c \in C$ to the PUF instance and generate the corresponding response $r \in R$.
(2) Reproducible: A PUF instance PUF_i is said to be *reproducible* if they generate the same or similar response with small margin of error when the same challenge is repeatedly applied to it. *Hamming distance* (HD) is typically used to compare the similarity of response sets for a given challenge. The measure of reproducibility is also called as *intra-distance* of the PUF since similarity is only measured for a given PUF instance and not across PUF instances. This property is also used to define the quality metric *Reliability*, which is described in Sect. 2.1.2.
(3) Uniqueness: Uniqueness property for a PUF instance is generally defined by how unique the PUF instance is in comparison to other instances of the same PUF design. Broadly, uniqueness metric can be defined for a given PUF variant, and more about it is discussed in quality metrics Sect. 2.1.2.
(4) Unclonable: A given PUF instance is described as *unclonable* if it is not physically clonable by the adversary or even the manufacturer. Due to this important feature, PUF primitives are advantageous for security applications, as they cannot be physically replicated by the manufacturer therefore, it is not necessary to have trust in the manufacturer.
(5) Unpredictable: Given a small collection of CRPs, it would be impossible for an adversary to construct an accurate mathematical model or hardware emulator corresponding to the PUF instance that would let the adversary to predict the response corresponding to an arbitrary challenge with high probability of success. Unfortunately, as we will see in later chapters, this *unpredictability* property of a given PUF instance does not hold in practice.
(6) One-way: A given PUF instance i is said to satisfy one-way property if it is hard to find the corresponding challenge from its response, e.g., for a given response r, it is hard to find c such that $r = PUF_i(c)$.
(7) Tamper-evident: A given PUF instance is said to be tamper-evident when physical attempts to alter a PUF instance result in a substantial and permanent change in the CRP behavior. Tamper-evident properties ensure that the PUF instance

under attack does not reveal its innate behaviors upon tampering, but instead transforms into a different PUF instance that differs significantly from the PUF instance under attack.

2.1.2 Quality Metrics of PUF

PUF has four important quality metrics. These are *uniqueness*, *reliability*, *uniformity* and *bit aliasing*. We discuss each of these parameters in detail.

a) **Reliability:** The reliability metric of a PUF is related to the reproducibility of the challenge-response behavior of individual PUF instances. This captures the reliability of the response value for the particular challenge **c**, which is typically done by collecting the response corresponding to the same challenge repeatedly. In fact, this metric can be used to determine the reliability of responses against a variety of environmental fluctuations, such as temperature or voltage. In order to compute reliability of the PUF instance, Hamming distance is used. Let the m-bit response be r and r' when the same challenge is applied twice where $r = (r_1, r_2, r_3 \ldots, r_m)$ and $r' = (r'_1, r'_2, r'_3 \ldots, r'_m)$, then hamming distance (HD) can be computed using the equation:

$$HD(r, r') = \sum_{i}^{m}(r_i \oplus r'_i) \qquad (2.1)$$

If r is the reference response and s the number of times the r' response is collected at different environmental/electrical conditions, then reliability (REL) of a PUF instance is computed by averaging the measured HD values (m denotes the number of response bits of the PUF variant). It is given by

$$\text{REL} = 1 - \frac{1}{s}\sum_{i=1}^{s}\frac{HD(r, r'_i)}{m} \times 100\% \qquad (2.2)$$

Ideally, the reliability of PUF should be 1, i.e., the same response value at different measurements.

b) **Uniqueness:** The Uniqueness metric of a PUF indicates the degree to which PUF instances belonging to the same PUF variant differ from each other. In order to determine the uniqueness of a pair of PUFs, Hamming distance (HD) is measured between them. For k PUF instances with response width m, uniqueness (UNQ) is computed as

$$\text{UNQ} = \frac{2}{k(k-1)}\sum_{i=1}^{k-1}\sum_{j=i+1}^{k}\frac{HD(r_i, r_j)}{m} \times 100\% \qquad (2.3)$$

where r_i, r_j represents the response for a given challenge **c** of PUF instance i and j respectively, where $1 \leq i, j \leq k$. The ideal value of PUF Uniqueness is 50%.

c) **Uniformity:** Uniformity is a measure of the distribution of "0s" and "1s" in the response bits to the random challenges. Ideally, the uniformity value should be 50%, i.e., equal distribution of "0s" and "1s". Hamming weight (HW) is used to measure the uniformity. Let $r = (r_1, r_2, r_3 \ldots, r_m)$ be the m-bit response of the given PUF instance for the challenge **c**, then uniformity (UNF) is given by

$$\text{UNF} = \frac{1}{m} \sum_{i=1}^{m} r_i \times 100\% \qquad (2.4)$$

d) **Bit-aliasing:** Bit-aliasing is a measure of the distribution of "0s" and "1s" of specific output bit across multiple instances. For the given k PUF instances with output width m, the lth binary bit aliasing (BIA) can be computed as

$$\text{BIA} = \frac{1}{k} \sum_{i=1}^{k} r_{i,l} \times 100\% \qquad (2.5)$$

where $r_{i,l}$ is lth binary bit of response generated by PUF instance i.

These four quality metrics are most commonly used to analyze the quality of PUF. However, there are also other metrics proposed by various authors. Readers are suggested to refer to the work [19, 20] for more details. Next, we will discuss different classifications of the PUF.

2.1.3 PUF Classification

Based on the size of their challenge-response space, PUFs are usually classified into the following classes [14]:

- **Strong PUF.** The CRP space of these PUFs are large enough to evade exhaustive characterization of the PUF circuit. This kind of PUFs are secure against challenge-replication-based attacks if the same challenge is not applied twice during protocol execution. The *Arbiter PUF*(APUF) [1] is a widely studied Strong PUF.
- **Weak PUF.** This type of PUFs has comparatively smaller number of CRPs, and susceptible to replication attack. The attacker with all the challenges will be able to emulate this PUF. Hence the responses of such PUFs are not exposed out of the hardware device. The examples of Weak PUF are SRAM PUF and *Ring Oscillator PUF* (ROPUF), because of the practical infeasibility of implementing ROPUFs with large number of challenge bits. For this reason, responses of these classes of PUFs are not allowed to leave the device. In the extreme case, a weak PUF might

not have any challenge and generate a constant instance-specific response, e.g., the *MECCA* PUF [10], which can be used for key generation and authentication.

Note that the above notion of "strength" of PUFs is not based on the security level of PUF. At current state of the art, a majority of strong PUFs are susceptible to modeling attacks though they are secured to emulation attacks.

Based on the way responses of PUF are generated, the PUF can be classified into following categories:

- **Delay PUF:** Delay PUFs are principally based on the inherent delay characteristics of the physical elements that vary from chip to chip. The difference in delay caused by an identically laid out pair of electrical paths, generates the response. Delay PUF can be strong PUF or Weak PUF based on its architecture. The widely studied examples of delay PUF are APUF and ROPUF.
- **Memory PUF:** This is based on the unpredictable power-up state values of bistable memory elements which occur due to random variations in transistor parameter values. The well-known example of Memory PUF is SRAM PUF [21].
- **Delay + Memory PUF:** Here the response is based on two possible stable states like memory PUFs, whereas the notion of delay PUF is utilized to build bistable delay components. The Bistable Ring PUF (BRPUF) [11] is an example of such a PUF, where bistable ring comprises an even number of inverting stages.

In spite of the fact that majority of the memory PUFs don't need challenges to generate responses and are considered as weak PUFs, the combination of some memory PUFs is used as strong PUF [12]. For example, consider SRAM cells which are organized in $l \times k$ grid to form $l \times k$ SRAM PUF. To make it use as strong PUF, the address can be considered to be the challenge bits of SRAM.

The rest of the chapter will focus on strong PUFs whose knowledge is crucial to understanding modeling attacks. Next, we will discuss well-studied popular silicon PUF called, *Arbiter PUF*.

2.2 Arbiter PUF (APUF)

The APUF consists of a cascade of several structurally identical two-port delay stages, with an arbiter (usually a latch or a flip-flop) at the end of the cascade. For each delay stage (say, the ith stage), which are path-swapping switches, connectivity from the input to output ports is controlled by a control bit (the ith challenge bit $c[i]$). The two input ports of the first stage are shorted, and an input pulse is allowed to propagate along the two delay paths. The exact paths followed by the two signals along the two paths depend on the challenge applied. Because of process variation effects, in spite of (ideally) identical layout, the signals along the two paths reach the arbiter after slightly different delays, resulting in either logic-0 or logic-1 output of the arbiter. Figure 2.2 shows an n-bit APUF circuit.

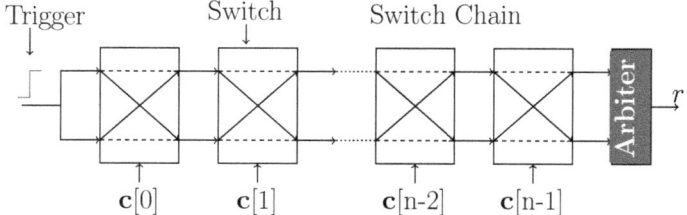

Fig. 2.2 An n-bit Arbiter PUF (APUF) [1]

For each n-bit challenge $\mathbf{c} \in \{0, 1\}^n$, it can be proved that the response of an APUF to for a given challenge $\mathbf{c}$ is given by

$$r = \begin{cases} 1 & \text{if } \Delta_\mathbf{c} < 0, \\ 0 & \text{otherwise.} \end{cases} \quad (2.6)$$

2.2.1 Delay Model of Arbiter PUF

It can be shown [1] that delay difference at the end of the cascaded switch stages, $\Delta_\mathbf{c}$ is given by $\Delta_\mathbf{c} = \mathbf{w}^T \Phi$. Here, $\mathbf{w}$ is known as the *weight vector*, and is a $(n + 1)$-dimensional vector of real numbers, with components dependent on the path delay, and Φ is the *parity vector* derivable from the given challenge $\mathbf{c}$, whose components are given by

$$\Phi[n] = 1 \quad \text{and} \quad \Phi[i] = \prod_{j=i}^{n-1}(1 - 2c[j]), i = 0, 1, \ldots n - 1 \quad (2.7)$$

This equation is deduced in Sect. 4.3. Interested readers can directly refer to the derivation.

2.3 PUF Composition Types

Although APUF has low hardware overhead and relatively simple structure, it is hardly usable in practice due to its vulnerability to modeling attacks which violates the "unpredictability" property of PUF [3]. However APUF was used as a popular building block to construct the composite PUFs. *PUF composition* is a PUF design paradigm that uses smaller PUFs as the building blocks for construction. The smaller PUFs are called as *constituent PUF or component PUF* and the resulting PUF as

2.3 PUF Composition Types

a *composite PUF*. There are a variety of ways to achieve the PUF composition. Currently, composite PUFs can be derived from three types of compositions: (a) combinational function based; (b) selection based, and (c) interpose based. Each of these composition types is described below, along with common examples of each category.

1. **Combinational Function-based Composition:** Let $f_i : \{0, 1\}^n \rightarrow \{0, 1\}$ be a ith instance of an n-bit constituent PUF, and $\mathbf{x}$ be the input challenge. Then, a combinational function-based composite PUF can be represented as

$$F(\mathbf{x}, k) := C(f_1(\mathbf{x}), \ldots, f_k(\mathbf{x})) \tag{2.8}$$

   where C is an combinational Boolean function. A well-known combination function-based composite PUF is XOR APUF [18] and Tribes APUF [13].

2. **Selection-based Composition:** Selection-based composite PUFs typically consist of 2^k individual PUF instances, among which the output PUF output is selected depending on the applied challenge, as the overall output of the composite PUF. Selection-based composite PUFs can be represented similar to combination based composite PUF, but the selection function typically involves other PUF instances. The output of such a PUF composition can be expressed as

$$F(\mathbf{x}, k) := MUX\left(f_{1,s}(\mathbf{x})||\ldots||f_{k,s}(\mathbf{x}); f_1(\mathbf{x}), \cdots, f_{2^k}(\mathbf{x})\right) \tag{2.9}$$

   where $MUX()$ denotes an 2^k:1 multiplexer, whose k-bit select input is given by given by $f_{1,s}(\mathbf{x})||\cdots||f_{k,s}(\mathbf{x})$, and its 2^k data inputs are given by $f_1(\mathbf{x})$ through $f_{2^k}(\mathbf{x})$. Note that each $f()$ mapping indicates an individual PUF instance. An example of selection-based composite PUF is the *Multiplexer PUF* (MPUF) [7]. Two variants of MPUF have also been proposed, called the *robost MPUF* (rMPUF) and *cryptanalytically robust MPUF* (cMPUF), respectively.

3. **Interpose-based Composition:** In Interpose-based Composition, the response bit(s) of one or multiple PUF instances are used to derive the challenge bits for a different set of one or more PUF instances, which ultimately produces the overall response. For simplicity, let us only consider the simplest composite PUF of this class—the Interpose PUF (iPUF) [3, 4, 8]. Mathematically, let $f_{up} : \{0, 1\}^p \rightarrow \{0, 1\}$ be an p-bit APUF with input challenge $\mathbf{x}$ and $f_{down} : \{0, 1\}^q \rightarrow \{0, 1\}$ be an $q(\geq p + 1)$-bit APUF, then the output response of an (p, q)-iPUF composition can be expressed as

$$F\left(f_{up}, f_{down}, i\right) := f_{down}(x_1, \ldots, x_{i-1}, f_{up}(\mathbf{x}), x_{i+1}, \ldots, x_q) \tag{2.10}$$

   where i is the interpose-bit position. Heterogeneous interpose PUF compositions consisting of PUF instances of multiple types have been described in [9].

Fig. 2.3 An x-XOR APUF

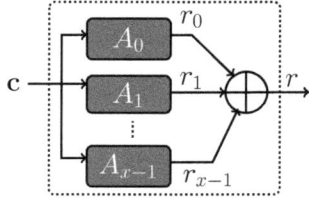

2.3.1 XOR PUF Composition

According to Yao's XOR lemma [13], the best non-linearity can be obtained by applying the XOR function as combination function. Hence, in XOR APUF composition, XOR is used as a combination function with arbiter PUF as constituent PUF. Mathematically, k-XOR APUF can be represented as

$$F_{xor}(\mathbf{x}, k) := \bigoplus_{i=1}^{k} f_i(\mathbf{x}) \qquad (2.11)$$

where f_i is ith instance of constituent PUF. Here, the input challenge to all constituent PUFs is the same. Upon generating the response from each constituent APUF, the responses are XORed to produce the single-bit final output response F_{xor}. In case of XOR arbiter PUF (XOR APUF), the constituent PUF employed is arbiter PUF. Figure 2.3 shows the structure of an k-XOR APUF. A slightly more complex PUF can be derived by controlling the input to each of the constituent PUFs rather than providing the same input challenge to all constituent PUF in XOR PUF composition. It can be achieved in two different ways: (1) to derive the different input n-bit challenge for each constituent PUF from a single n-bit challenge. This method is called as input challenge transformation. (2) to apply a completely different n-bit challenge to each of the constituent PUFs. The popular PUF examples of above-mentioned types are *Lightweight Secure PUF* (LSPUF) and *Pseudo-random challenge driven XOR PUF* (PC-XOR PUF).

2.3.1.1 Lightweight Secure PUF (LSPUF)

The LSPUF was designed to enhance the security over the XOR APUF [6]. The architectural enhancement involves the derivation of sub-challenge c from master challenge $\mathbf{C}$ for each constituent APUF. Consider Q constituent APUFs, with numbers starting at the top and going down and this derivation can be achieved in two steps: rotation and application of transformation rule. Initially, for the kth APUF instance, the n-bit master challenge $\mathbf{C}$ is rotated by k positions to derive: $\mathbf{d} = \text{rotate}(\mathbf{C}, \mathbf{k})$, and then the following transformation is applied:

2.3 PUF Composition Types

Fig. 2.4 An m-output Lightweight Secure PUF (LSPUF) [6]

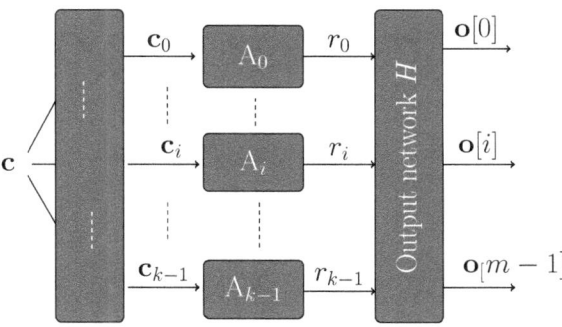

$$c_{\frac{N+i+1}{2}} = d_i \quad \text{for i =1}$$
$$c_{\frac{i+1}{2}} = d_i \oplus d_{i+1} \quad \text{for } i = 1, 3, 5, \ldots N - 1$$
$$c_{\frac{N+i+2}{2}} = d_i \oplus d_{i+1} \quad \text{for } i = 2, 4, 6, \ldots N - 2$$

unlike XOR APUFs, LSPUF is a multi-bit output PUF (say, m output bits), where each output bit is generated by XOR-ing a fixed number (say, x) of outputs of a set of APUFs (Q total APUFs), according to the formula:

$$\mathbf{o}[i] = \bigoplus_{j=0}^{x-1} r_{((i+s+j) \mod Q)} \quad (2.12)$$

Figure 2.4 shows an example LSPUF with m-output bits, and Q constituent APUFs.

2.3.1.2 Pseudo-random Challenge Driven Input XOR APUF (PCXOR APUF)

Yu et al. [16] proposed PCXOR APUF where each constituent APUF is fed with entirely different challenges. These challenges are generated using pseudo-random number generators (PRNGs). Figure 2.5 shows the 64-bit 4-PCXOR APUF. A 256-bit challenge vector was generated with PRNG and sliced into four 64-bit challenges. Each challenge was assigned constituent APUF.

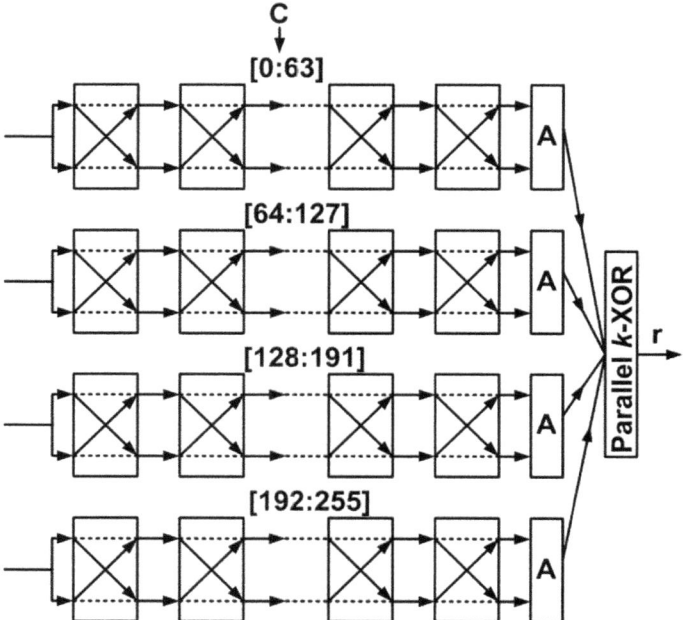

Fig. 2.5 A 64-bit 4-PCXOR APUF [16]

2.3.1.3 Double Arbiter PUF (DAPUF)

k-l DAPUF was designed to address the problem of uniqueness of arbiter PUF in FPGA [15] where k denotes the number of arbiter chains and l denotes the number of output responses. Although DAPUF uses XOR function for combination, its building block PUF is slightly different from the classic APUF. Here, all top path signals from k arbiter PUF compete with all bottom path signals. For example, unlike an 2-XOR APUF where the arbiter receives the input from top to bottom signal of same APUF chain, 2-1 DAPUF involves top path of first and second APUF delay chain is connected to arbiter-1, and bottom path of the first and second APUF delay chains connected to arbiter-2. Further, the responses from arbiter-1 and arbiter-2 are XOR-ed to get the single response. Figure 2.6 describes the structure of 2-1 DAPUF.

2.3.2 Tribes PUF Composition

Recently the *Tribes Combined Arbiter PUF* (Tribes APUF) has been proposed [13], whose combinational Boolean function is the *Tribes function*. The Tribes function has two parameters k and b, and is given by

2.3 PUF Composition Types

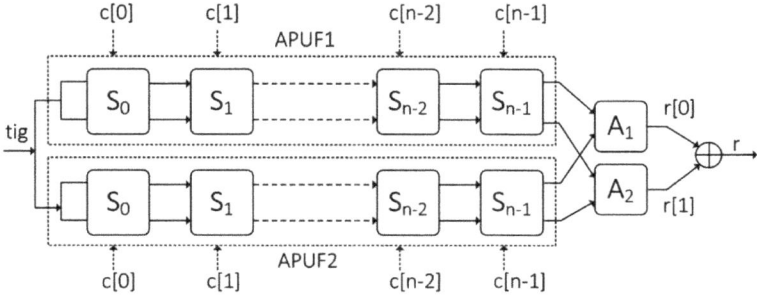

Fig. 2.6 2-1 Double Arbiter PUF (DAPUF) [15]

$$T(x_1, \ldots, x_k) := (x_1 \wedge \cdots \wedge x_b) \vee \cdots \vee (x_{k-b+1} \wedge \cdots \wedge x_k) \quad (2.13)$$

where k denotes the number of variables and parameter b is set to a certain quantity a slightly less than $\log_2 k$ [17]. In Tribes function, the k inputs are divided into $\frac{k}{b}$ blocks each of size b, and it's output is set to be one if and only if at least one of the blocks consists entirely of ones. In order for the Tribes function to be balanced, k and b should be defined in such a way that $(1 - 2^{-b})^{k/b} \approx \frac{1}{2}$ [17]. Also, please note that balanced Tribes function can be defined for only a few input lengths, since in that case k is required to be a multiple of b. The Tribes APUF replaces the individual Boolean variables in the above expression with outputs of individual APUFs. For example, the Tribes APUF with $k = 4$, $b = 2$ can be represented as

$$F_{Tribes}(\mathbf{x}, 4) := (f_1(\mathbf{x}) \wedge f_2(\mathbf{x})) \vee (f_3(\mathbf{x}) \wedge f_4(\mathbf{x})) \quad (2.14)$$

2.3.3 Multiplexer PUF (MPUF) and Its Variants

Multiplexer PUF (MPUF) employs selection-based PUF composition. The primary goal of MPUF and its variants was to reach robustness to cryptanalysis and modeling attack comparable to XOR PUF, while being much more reliable than the XOR APUF. An (n, k)-MPUF employs an $2^k:1$ multiplexer, with the k select lines of the multiplexer being connected to the outputs of k APUFs, and the data lines of the multiplexer being connected to 2^k APUFs, with all AUFs are input the same n-bit challenge. Figure 2.7 shows the architecture of an (n, k)-MPUF. The two other variants were also proposed, viz., cMPUF (robust against cryptanalysis) and rMPUF (robust against reliability-based modeling attack). The rMPUF uses 2^{k-1} APUFs for data selection using a multiplexer tree, while the cMPUF uses only 2^{k-1} data APUFs, with the other 2^{k-1} data inputs being complements of them.

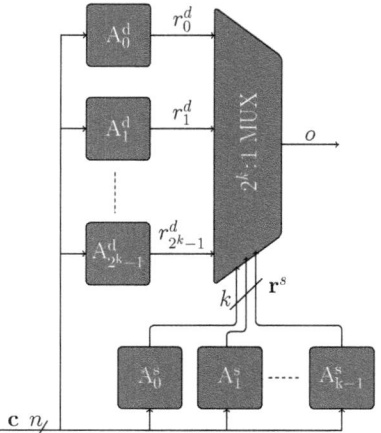

Fig. 2.7 Architecture of an (n, k)-MPUF [7]

2.3.4 Interpose PUF (IPUF)

Interpose PUF (IPUF) employs Interpose-based PUF composition and is the latest addition to the repertoire of robust PUFs. The idea is to interpose the response of an x-XOR APUF (with n challenge bits) to an applied challenge, between two challenge bits of an y-XOR APUF (with $(n + 1)$ challenge bits, the remaining challenge bits being the same as the x-XOR PUF), to get an (x, y)-IPUF. Figure 2.8 shows the structure of an (x, y)-IPUF, with n-bit challenge.

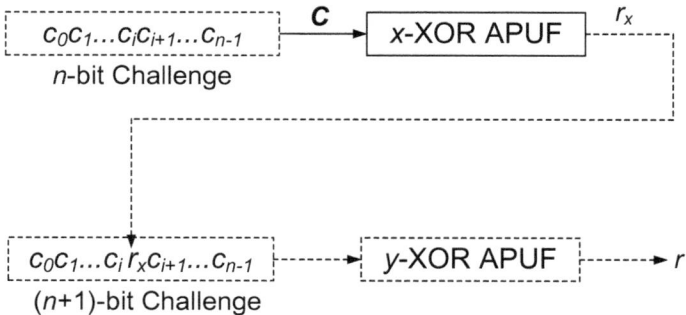

Fig. 2.8 Architecture of an n-bit (x, y)-IPUF [8]. Only the n-bit challenge c and the 1-bit response r are accessible as circuit ports

References

1. Lim, D. (2004). *Extracting secret keys from integrated circuits*. Master's thesis, Massachusetts Institute of Technology, U.S.A.
2. Gassend, B., Clarke, D., van Dijk, M., & Devadas, S. (2002). Silicon physical random functions. In *Proceedings of the 9th ACM Conference on Computer and Communications Security, Ser. CCS'02* (pp. 148–160).
3. Santikellur, P., Bhattacharyay, A., & Chakraborty, R. S. (2019). Deep learning based model building attacks on arbiter PUF compositions. *IACR Cryptology Eprint Archive Journal, 2019*, 566.
4. Wisiol, N., Mühl, C., Pirnay, N., Nguyen, P. H., Margraf, M., Seifert, J.-P., van Dijk, M., & Rührmair, U. (2019). Splitting the interpose PUF: A novel modeling attack strategy. *Cryptology ePrint Archive*, Report 2019/1473, https://eprint.iacr.org/2019/1473.
5. Rührmair, U., Sehnke, F., Sölter, J., Dror, G., Devadas, S., & Schmidhuber, J. (2010). Modeling attacks on physical unclonable functions. *Proceedings of the ACM Conference on Computer and Communications Security, Ser. CCS'10*, 237–249.
6. Majzoobi, M., Koushanfar, F., & Potkonjak, M. (2008). Lightweight secure PUFs. *Proceedings of the IEEE/ACM International Conference on Computer-Aided Design, Ser. ICCAD'08*, 670–673.
7. Sahoo, D. P., Mukhopadhyay, D., Chakraborty, R. S., & Nguyen, P. H. (2018). A multiplexer-based arbiter PUF composition with enhanced reliability and security. *IEEE Transactions on Computers, 67*(3), 403–417.
8. Nguyen, P. H., Sahoo, D. P., Jin, C., Mahmood, K., Rührmair, U., & van Dijk, M. (2018). The interpose PUF: Secure PUF design against state-of-the-art machine learning attacks. Accessed May 2018, from https://eprint.iacr.org/2018/350.
9. Sahoo, D. P., Saha, S., Mukhopadhyay, D., Chakraborty, R. S., & Kapoor, H. (2014). Composite PUF: A new design paradigm for physically unclonable functions on FPGA. In *Proceedings of the IEEE International Symposium on Hardware-Oriented Security and Trust, Ser. HOST'14* (pp. 50–55).
10. Krishna, A. R., Narasimhan, S., Wang, X., & Bhunia, S. (2011). MECCA: A robust low-overhead PUF using embedded memory array. In B. Preneel & T. Takagi (Eds.), *Cryptographic hardware and embedded systems* (pp. 407–420). Nara, Japan: CHES 2011: 13th International Workshop.
11. Chen, Q., Csaba, G., Lugli, P., Schlichtmann, U., & Rührmair, U. (2011). The bistable ring PUF: A new architecture for strong physical unclonable functions. In *Proceedings of IEEE International Symposium on Hardware-Oriented Security and Trust (HOST)* (pp. 134–141).
12. Holcomb, D. E., & Fu, K. (2014). Bitline PUF: Building native challenge-response PUF capability into any SRAM. In *Proceedings of Cryptographic Hardware and Embedded Systems (CHES)* (pp. 510–526).
13. Ganji, F., Tajik, S., Stauss, P., Seifert, J.-P., Forte, D., & Tehranipoor, M. (2019). Rock'n'roll PUFs: Crafting provably secure PUFs from less secure ones. In *Proceedings of 8th International Workshop on Security Proofs for Embedded Systems, 11*, 33–48.
14. Maes, R. (2013). *Physically unclonable functions - constructions*. Properties and Applications: Springer.
15. Machida, T., Yamamoto, D., Iwamoto, M., & Sakiyama, K. (2015). A new arbiter PUF for enhancing unpredictability on FPGA. *The Scientific World Journal, 2015*.
16. Yu, M. -D., Hiller, M., Delvaux, J., Sowell, R., Devadas, S., & Verbauwhede, I. (2016). A lockdown technique to prevent machine learning on PUFs for lightweight authentication. *IEEE Transactions on Multi-Scale Computing Systems, 2*(3), 146–159.
17. O'Donnell, R. (2004). Hardness amplification within NP. *Journal of Computer and System Sciences, 69*(1), 68–94.
18. Suh, G. E., & Devadas, S. (2007). Physical unclonable functions for device authentication and secret key generation. In *Proceedings of the Design Automation Conference, Ser. DAC'07* (pp. 9–14)

19. Pehl, M., Punnakkal, A. R., Hiller, M., & Graeb, H. (2014). Advanced performance metrics for physical unclonable functions. In *International Symposium on Integrated Circuits (ISIC)* (pp. 136–139).
20. Wilde, F., Gammel, B. M., & Pehl, M. (2018). Spatial correlation analysis on physical unclonable functions. *IEEE Transactions on Information Forensics and Security, 13*(6), 1468–1480.
21. Guajardo, J., Kumar, S. S., Schrijen, G. -J., & Tuyls, P. (2007). FPGA intrinsic PUFs and their use for IP protection. In *International Workshop on Cryptographic Hardware and Embedded Systems* (pp. 63–80). Springer.

Chapter 3
Machine-Learning Basics

3.1 Introduction

In Chap. 1, we discussed how machine-learning algorithms differ from traditional algorithms. This chapter will introduce some machine-learning terminologies as well as some details about the working of ML algorithms.

Consider the machine learning-based spam detection system which detects the email is spam or not. Before we make it work, the system has to be trained with large set of example data. A collection of example data is called *dataset*. The way the system learns to construct an approximation prototype from the dataset is determined by the machine-learning algorithm. In our example, the method to derive the logic of spam detection using the available dataset is determined by machine-learning algorithm. There are several ML algorithms proposed to learn from the data. The output of the ML algorithm is called model. The ML model represents the learned system.

In our spam detection system, the dataset is comprised of multiple samples, each with an email text and whether it is spam or not. In this case, the text of the individual emails in the training set, can be considered the "feature", and the corresponding "yes" or "no" answer of whether the email is spam or not can be considered as the training sample "label". The dataset is typically divided into three parts: the training dataset, the validation dataset, and the test dataset. This corresponds with the way the ML model is built: the training phase, the validation/testing phase, and the application phase. As part of the training phase, the training dataset is used for learning the model. The validation phase evaluates the learning performance of the model using validation dataset and the application phase involves deriving inferences from the newly developed model using test dataset. The test dataset should only be exposed during the application phase and not for the first two phases.

The dataset which contains both features and corresponding labels (for a sample) is called a "labeled dataset". However, it is not always the case that the dataset consists of features and labels. In the real world, data are not always labeled. Hence, sometimes, only features are present in the dataset. such dataset are called "unlabeled dataset".

3.2 Machine-Learning Classification

Depending on the dataset and the kind of problem statement, machine learning can be divided into three categories: supervised learning, unsupervised learning, and reinforcement learning.

(1) **Supervised learning:** In supervised learning, the ML model is trained under supervision, i.e., the actual output guides the expected output of the model at each step of the training process. For example, spam detection is trained using supervised learning, since labels ("yes" or "no") are used by learning algorithm to determine if email content is spam or not. As you might have guessed now, supervised learning is associated with labeled datasets.

Let us consider input features as **x** and output labels as y. Mathematically, mapping between the input features and labels can be represented as

$$y = f(\mathbf{x}) \tag{3.1}$$

Supervised learning aims to learn the function f using **x** and y. We can use supervised learning for two types of problems: classification and regression. For classification problems, y is discrete (categorical), while for regression problems, y is continuous. The y values in spam detection are categorical (e.g., "yes" or "no"). Hence, it's a classification problem. Examples of regression problems include predicting house prices, salaries, etc.

(2) **Unsupervised learning:** The objective of unsupervised learning can be to discover the underlying pattern in an unlabeled dataset. Contrary to supervised learning, which aims to make right predictions for the new data, unsupervised learning seeks to glean insight from the unlabeled data. Anomaly detection and recommendation systems are popular applications of unsupervised learning. Two of the most important problems that can be solved using unsupervised learning are *clustering* and *dimensionality reduction*.

   (a) **Clustering:** In essence, clustering is a classification task on unlabeled datasets. As its name suggests, the goal of clustering is to find natural groupings within data. Data points that have similar characteristics will belong to the same cluster. There have been several algorithms designed to cluster data based on their similarity definition. *K-means* is one of the most popular clustering algorithms where n total data points are grouped into k clusters in which the similarity of points in each cluster is determined by the Euclidean distance.

   (b) **Dimensionality reduction:** It is one of the most powerful techniques in ML to reduce the dimensions of a dataset. A dataset usually contains sparse or correlated data, so large dimension dataset have a serious computational impact. These techniques scale down the number of features in a dataset without compromising the integrity of the data. These techniques are generally applied to the preprocessing of datasets before modeling. *Principal*

Component Analysis(PCA) [11] and *Autoencoders* [10] are popular dimensionality reduction algorithms.

(3) **Reinforcement learning:** Reinforcement learning (RL) is an entirely different paradigm compared to supervised and unsupervised algorithms. RL is based on the idea of learning by interacting with the environment. The goal of the RL problem is to map situations to actions so as to maximize the numerical reward. RL is also synonymous with the natural method of learning. For example, the child learns to walk by observing the environment and attempting different movements while maintaining balance, and the child is punished when they fall and rewarded when they move forward. In addition, there are many other examples, such as learning to ride a bicycle, or engaging in a conversation where we are very aware of how our environment will respond based upon what we do and we attempt to influence what happens by how we act. RL is a much more focused, goal-oriented approach than the other two methods.

Most machine-learning algorithms used in hardware security fall into the unsupervised or supervised category. In the following section, we will discuss some of the most commonly used supervised and unsupervised ML algorithms in hardware security.

3.3 Supervised ML Algorithms

3.3.1 Support Vector Machines

Support vector machines (SVM) was proposed by Boser, Guyen and Vapnik [12] based on statistical learning theory to classify the binary data. SVM is an excellent method for computational learning because of several outstanding advantages. It can be applied to classification and regression problems and is appropriate for linear and non-linear data processing. Additionally, it has special generalization capabilities particularly for smaller datasets and does not suffer from local minimum problem (Fig. 3.1). Before understanding the SVM, we introduce you few concepts. Consider a set of 2D data points belonging to two classes, as shown in Fig. 3.2a. The data is called *linearly separable* if the two classes can be separated by a straight line so that one class will be on one side and the other class on the other. Figure 3.2a, b shows the set of data points that are linearly separable and linearly inseparable. The line that separates two classes is called the *separating hyperplane*. In two dimensions, the separating hyperplane is a line, and in three dimensions, it is a plane. Hyperplane is considered the higher dimensional equivalent of plane in three dimensions. Consequently, in the case of n-dimensional linearly separable data, a hyperplane is a decision boundary that will separate the two classes.

It follows that there can be many hyperplanes that can classify the data, but which one is the best?. The best hyperplane is the one that maintains the maximum distance

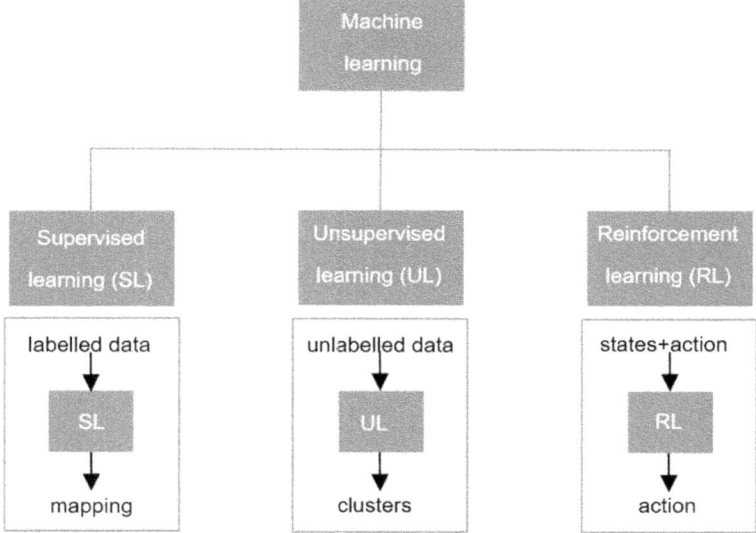

Fig. 3.1 Machine-learning classification

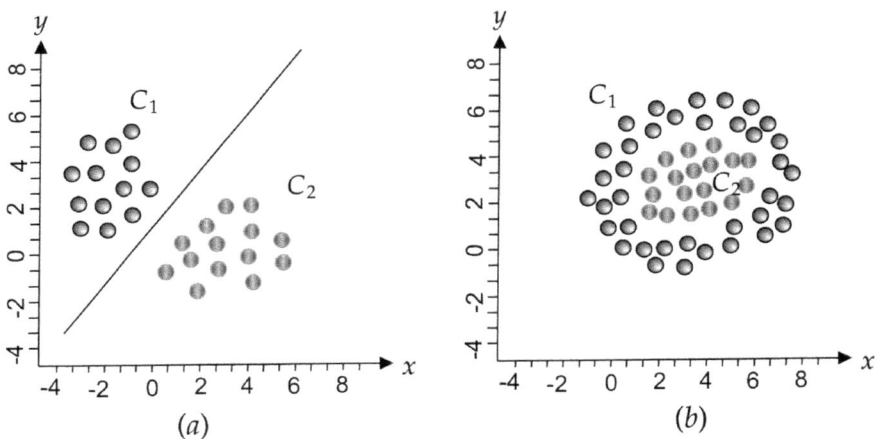

Fig. 3.2 a Linearly separable data points b linearly inseparable data points

from its nearest data points of both classes. This distance is called *margin* and the nearest data points to the hyperplane are called *support vectors*. In order to determine the best hyperplane, SVM models this problem as an optimization problem where the objective is to determine the hyperplane with the maximum margin between support vectors.

Figure 3.3 shows the support vectors and margin for the considered data points. Consider the data point x_1 in Fig. 3.3, the distance between x_1 and the hyperplane

3.3 Supervised ML Algorithms

Fig. 3.3 An example of SVM-based classification

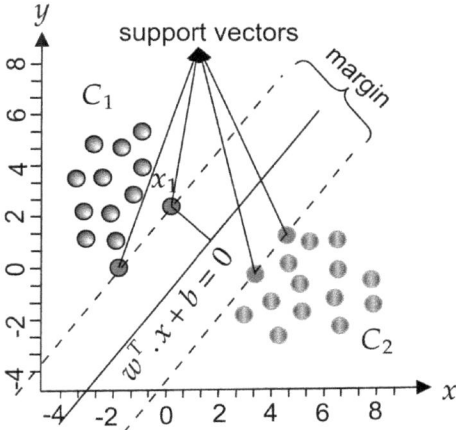

is $\frac{\mathbf{w}^T \cdot \mathbf{x} + b}{||w||}$. Transforming the label from 0, 1 to $-1, +1$ (bipolar encoding) helps to simplify the mathematical steps. The goal of finding the hyperplane with maximum margin from support vectors that are closest to the hyperplane can be expressed as

$$\underset{\mathbf{w},b}{\mathrm{argmax}} \left\{ \min_n (\mathrm{label}.(\boldsymbol{w}^T \cdot \boldsymbol{x} + b) \cdot \frac{1}{||w||} \right\} \quad (3.2)$$

This can be converted to constrained optimization problem with setting (label.$(\boldsymbol{w}^T \cdot \boldsymbol{x} + b)$) to 1.0 or greater. Such constrained problems can be solved using *Lagrange multipliers*. Using *Lagrange multipliers* (α), the optimization function can be formulated as

$$\max_\alpha \left[\sum_{i=1}^m \alpha - \frac{1}{2} \sum_{i,j=1}^m \mathrm{label}^{(i)} \cdot \mathrm{label}^{(j)} \cdot \alpha_i \cdot \alpha_j \cdot \langle x^{(i)}, x^{(j)} \rangle \right] \quad (3.3)$$

subject to the following constraints:

$$c \geq \alpha \geq 0 \text{ and } \sum_{i=1}^m \alpha_i \cdot \mathrm{label}^{(i)}$$

The variable c is called the slack variable because it measures the trade-off between making the margin maximum and correct classification of data points.

3.3.1.1 Kernel Functions for Non-linear Classification

For data that is linearly separable, Eq. (3.3) is accurate. However, the *kernel functions* [12] are effective for complex data patterns that require non-linear separating surface. Non-linear classification by SVM relies on the idea that any arbitrary dataset can be linearly separable if the input features are mapped into high-dimensional features. This conversion is performed by kernel functions. High-dimensional feature mapping-based classification can be achieved by replacing the dot product computation of features in Eq. (3.3) (i.e., $\langle x^{(i)}, x^{(j)} \rangle$) with kernel functions. This replacing method is called the *kernel-trick*. The popular kernel function often used is *radial-basis function*, which is given by

$$K(x, y) = \exp\left\{\left(\frac{-||x-y||^2}{2\sigma^2}\right)\right\} \quad (3.4)$$

SVM is not the only machine-learning algorithm to use kernel functions. Other ML algorithms also make use of them. *Polynomial kernels* and *sigmoid kernels* are some of the more popular kernel functions.

3.3.2 Logistic Regression

In general, linear regression is a method of finding the best line that fits the data points. it uses optimization like *gradient descent* to find the best fit parameters. However, linear regression is not suitable for classification because it does not output the probabilities of the class to which it could belong. By defining the *logistic function* or *sigmoid function*, logistic regression achieves classification using the same optimization methods as linear regression.

The logistic function or sigmoid function (σ) is defined by

$$\sigma(z) = \frac{1}{1+e^{-z}} \quad (3.5)$$

The sigmoid function maps the predicted values (quantities) to probabilities. For any real input value z, the output of sigmoid function is bounded between 0 and 1. Hence, it is also known as *squashing function*. Figure 3.4 shows the plot of sigmoid function.

S-shaped sigmoidal curve intersects $z = 0$ in the middle of the ordinate y (i.e., $y = 0.5$). Also,

$$y \lesseqgtr \begin{cases} < 0.5 & \text{if } z < 0, \\ > 0.5 & \text{if } z > 0 \end{cases} \quad (3.6)$$

3.3 Supervised ML Algorithms

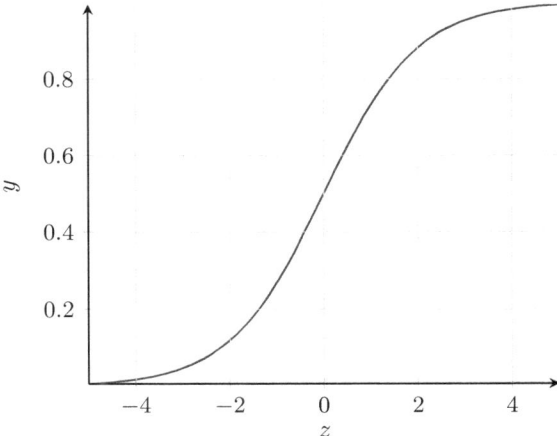

Fig. 3.4 Sigmoid function

In addition to these interesting properties, it is also differentiable, has a simple derivative form, and most importantly can be used to model conditional probability distributions.

The linear approach of modeling the relationship between dependent and independent variable (linear regression) is given by

$$\hat{z} = w_0 x_0 + w_1 x_1 + \cdots + w_n x_n$$

where x_i are inputs and w_i are parameters of the model. The prediction are converted to probabilities between 0 and 1 by feeding the output of linear modeling equation as input to sigmoid function.

$$P(z = 1) = \frac{1}{1 + \exp\left\{-(w_0 x_0 + w_1 x_1 + \cdots + w_n x_n)\right\}} \quad (3.7)$$

To achieve successful classification, parameter values w_i are tuned using optimization algorithms such as gradient descent.

The linear logistic regression has been more common and available in popular Ml frameworks like *sci-kit learn*, but logistic regression-based modeling can be used for a wide variety of problems having varying levels of complexity. some of the powerful logistic regression-based modeling techniques [9] are *multiple logistic regression, polynomial logistic regression* and *multinomial logistic regression*. The interested reader is referred to [9] for detailed insights on these techniques.

3.3.3 Artificial Neural Networks

Many cognitive scientists study the brain in order to understand how it works, so they stimulate the natural neural networks. Engineering development of neural networks is based on the belief that building a computational model inspired by the brain could be better than current computers, since the brain has incredible information processing capabilities. In many cases, traditional systems fail to perform tasks such as object recognition and speech recognition, but if built on machines, these solutions have great economic value. Hence, understanding the functioning of the brain may help in devising the methods which can solve these tasks.

The human brain is composed of a network of interconnected processing units called *neurons* that use biochemical reactions to send, receive, and process information. There are 100 billion neurons in an adult brain, and each neuron is connected by 10000 connections, which are called *synapses*. This makes it the most complex system and very different from current computing systems which have typically one to eight processors. Furthermore, memory in the brain is associated with synapses between two neurons and it is distributed throughout the network, whereas in traditional systems memory is a separate unit from the processor. However, it is believed that brain processing units are simpler and slower than conventional processors. There are many neuron-based learning models available in the literature. Next, we will examine a popular one known as *perceptrons*.

3.3.3.1 Perceptrons

The perceptron model mimics the simple computational model of a neuron. Similar to natural neurons, which have multiple inputs (*dendrites*) and a single output (*axon*), perceptrons also have multiple inputs and a single output. The synapses between dendrites and axons consist of modulated electrical signals of different strengths. The neuron only fires when the aggregate strength of inputs exceeds a threshold value. Accordingly, the perceptron represents the input signal strength as vector of weight values plus a distinguished value b, called bias. The output of a perceptron is the weighted sum of the inputs plus bias, piped through a step function.

As an example, let x_1 and x_2 be the inputs to the perceptron, and let $\mathbf{w} = [w_1, w_2]$ and b be *weights* and *biases*, respectively. Then the output of the perceptron (y) will be given by

$$y = \begin{cases} 1 & \text{if } w_1 x_1 + w_2 x_2 + b > 0, \\ 0 & \text{otherwise} \end{cases} \quad (3.8)$$

Figure 3.5 shows the perceptron model of neuron. $\mathbf{w}$ and b are called parameters of the perceptron. The step function is the simplest of the group of functions called *activation functions*. Equation (3.8) can be expressed as a dot product as follows:

3.3 Supervised ML Algorithms

Fig. 3.5 Perceptron model of neuron

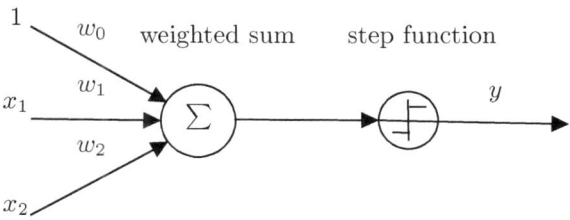

$$y = \begin{cases} 1 & \text{if } \mathbf{w} \cdot \mathbf{x} + b > 0 \\ 0 & \text{otherwise} \end{cases} \tag{3.9}$$

The *perceptron learning rule* was developed for supervised binary classification. It adjusts the weight and bias to make better performance in classification. However, perceptron algorithms are linear classifiers that can only discriminate when the data points are linearly separable. Next, we will explore a generalized and more complex perceptron-based model that can work for non-linear and multi-class problems.

3.3.3.2 Feedforward Neural Networks

According to our previous discussion, a perceptron-based neuron model is a combination of aggregation and activation. As we have discussed, a perceptron-based neuron model is represented as a combination of aggregation and activation function. Feedforward neural networks can also be called multi-layer perceptrons, where the neurons are organized in layers and interconnected in feedforward fashion.

Feedforward neural networks have three types of layers: *input layer*, *hidden layer(s)* and *output layer*. The input layer receives inputs, and the output layer produces the desired output. The layer(s) in between the input layer and output layer are called hidden layer(s). Figure 3.6 shows the feedforward neural networks with three hidden layers.

The popular activation functions [4] that are used for neurons in the network are *Sigmoid*, *Tanh*, *ReLu*, and *SELU*. Typically, ReLu is applied to hidden layer neurons, while *Sigmoid* or *Softmax* are applied to output neurons based on either binary or multi-class classification. Each of them are defined below.

Typical feedforward neural networks predict in the following way: input data is received at the input layer, propagated through hidden layers in forward direction. For each hidden neuron, intermediate output values are computed by combining aggregation and activation, and those values are then passed to the neurons of the next layer until the final output is determined.

Learning in neural networks means adjusting their parameters in a way that will result in a better classification performance. This is accomplished by using a optimization algorithm such as *Stochastic Gradient Descent* (SGD) [8] and *Adam optimizer* [2]. Prior to training, the parameters of the neural network are first initialized

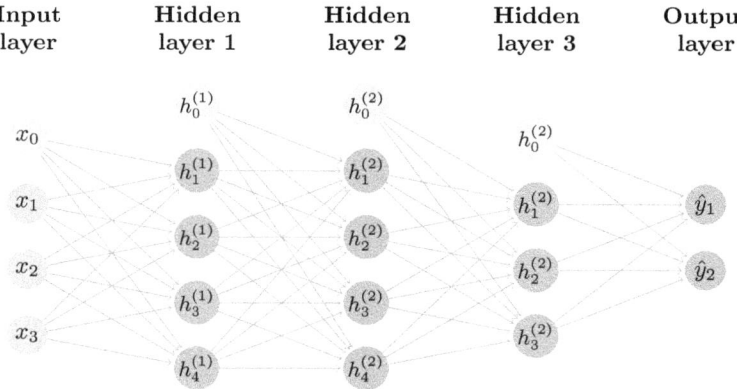

Fig. 3.6 Feedforward neural networks with three hidden layers

either randomly or using several different initialization methods, such as *Xavier* [1] and *He* [7]. An efficient method to compute the gradients in neural networks is accomplished through *backpropagation* method. It consists of two phases:

1. Forward pass: During this phase, the inputs are propagated from the input layer to the output layer in forward direction, and all intermediate computations in forward phase are stored. Using the *loss function*, the error between the actual output and the computed output is computed.
2. Backward pass: During this phase, the parameters of network are adjusted in the reverse order, starting at the output layer, working its way down to the input layer. By using *chain rule*, gradient of loss term with respect to each parameter is computed. Further, these gradients are used by optimization algorithms to update the weights.

It has been shown that feedforward networks with as few as one hidden layer are indeed capable of universal approximation in a very precise and satisfactory sense [6]. However, the number of parameters and required dataset size increases exponentially as the number of neurons increases in a single hidden layer. Single hidden layer training is also associated with problems such as vanishing and exploding gradients and local minima. Later, it was found that complex problems can be generalized better when more layers are used along with effective optimizing techniques. In the evolution of deep learning, feedforward neural networks that have more than one layer of hidden units are called deep neural networks (DNN) [4, 5]. The "deep" is associated with "representation learning", which allows a DNN architecture to automatically discover the hierarchical representations necessary for classification [3].

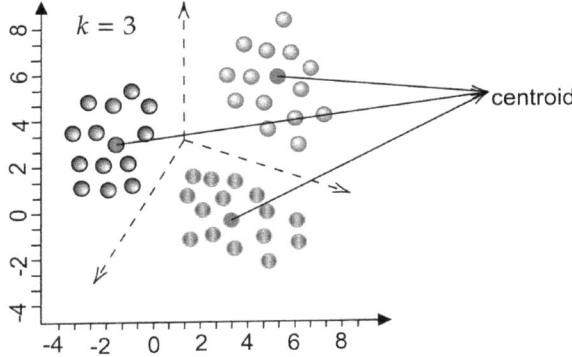

Fig. 3.7 An example clustering of K-means algorithm

3.4 Unsupervised ML Algorithms

3.4.1 K-means Clustering

K-means clustering is a technique that groups data points with similar characteristics into k number of clusters. The center of each cluster is the mean value of its data points. Since K-means clustering groups the data points by using no target labels, it can be viewed as k-class unsupervised classification. Figure 3.7 shows an example output by K-means algorithm.

Cluster formation by K-means algorithms is achieved iteratively. In the initial step, k random data points are chosen from the set of data points that comprise k clusters and used as the center (centroid) of the cluster. Next, the data points are assigned to the cluster whose centroid is closest to the data point, and centroids are updated by taking the mean of all data points in the cluster. As a result, data points within a cluster are closer together while keeping maximum distance between clusters.

The pseudo-code of the K-means algorithm is shown in Algorithm 1.

Algorithm 1: K-means algorithm

Input : A set of data points $\{x_1, x_2, x_3 \ldots, x_n\}$
K-number of desired clusters
Output:: K clusters
Randomly initialize cluster centroids $\mu_1, \mu_2, \mu_3 \ldots \mu_k$
repeat
    1. Assign the each data point x_i to the closest centroid μ_j (closest cluster)
    2. Recompute the centroid of each cluster
until *no change in centroid position*;

Computing the nearest centroid is based on the distance measure. While the K-means algorithm works with any distance measure, the Euclidean distance is most commonly used.

References

1. Glorot, X., & Bengio, Y. (2010). Understanding the difficulty of training deep feedforward neural networks. In *Proceedings of the International Conference on Artificial Intelligence and Statistics, Ser. AISTATS'10* (pp. 249–256).
2. Kingma, D. P., & Ba, J. (2015). Adam: A method for stochastic optimization. In *Proceedings of the International Conference on Learning Representations, Ser. ICLR'15*.
3. LeCun, Y., Bengio, Y., & Hinton, G. (2015). Deep learning. *Nature, 521*, 436–444.
4. Goodfellow, I., Bengio, Y., & Courville, A. (2016). *Deep Learning*. MIT Press, http://www.deeplearningbook.org.
5. Hinton, G., et al. (2012). Deep neural networks for acoustic modeling in speech recognition. *IEEE Signal Processing Magazine, 29*.
6. Hornik, K. (1991). Approximation capabilities of multilayer feedforward networks. *Neural Networks, 4*(2), 251–257.
7. He, K., Zhang, X., Ren, S., & Sun, J. (2015). Delving deep into rectifiers: Surpassing human-level performance on imagenet classification. In *Proceedings of the IEEE international conference on computer vision* (pp. 1026–1034).
8. Robbins, H., & Monro, S. (1951). A stochastic approximation method. *The Annals of Mathematical Statistics,* 400–407.
9. Hosmer Jr, D. W., Lemeshow, S., & Sturdivant, R. X. (2013). *Applied logistic regression* (Vol. 398) Wiley.
10. Hinton, G. E., & Salakhutdinov, R. R. (2006). Reducing the dimensionality of data with neural networks. *Science, 313*(5786), 504–507.
11. Jolliffe, I. T. (1986). Principal components in regression analysis. *Principal Component Analysis* (pp. 129–155). Springer.
12. Boser, B. E., Guyon, I. M., & Vapnik, V. N. (1992). A training algorithm for optimal margin classifiers. In *Proceedings of the Fifth Annual Workshop on Computational Learning Theory* (pp. 144–152).

Chapter 4
Modeling Attacks on PUF

4.1 Introduction

Modeling attacks are considered to be the greatest threat against strong PUF implementations, and wide-ranging intensive research is currently underway to develop newer attacks. In these attacks, typically a small fraction of the CRP dataset of a particular PUF instance is made available to the adversary, based on which she aims to build an accurate computational model of the PUF instance, capable of predicting the response for an arbitrary challenge with high probability of success. A successful attempt would compromise the PUF and the protocols that are built on it.

In order to illustrate this clearly, let us look at a strong PUF instance with n-bit input challenges and m-bit output response, following the Boolean mapping $f: \{0, 1\}^n \rightarrow \{0, 1\}^m$. The challenge space of this PUF instance is 2^n. To construct the computational model, t number of CRPs are used where $t \ll 2^n$. Based on effective model construction, any response for the corresponding challenge from remaining CRP space, i.e., $(2^n - t)$ can be derived. The number of CRPs needed to launch the successful modeling attack on a PUF varies depending on the robustness of PUF against modeling attacks. Typically, machine learning (ML) techniques are considered as natural choice for modeling attacks.

Modeling attacks using ML can be realized using supervised learning methods. The CRP dataset is partitioned into training set and testing set. The training data for a supervised learning algorithm consists of challenge-response pairs. During the training process, the supervised ML algorithm attempts to identify the relationship between challenges and responses. The learnt model can then be used to predict responses to unseen challenges. Figure 4.1 shows the process of machine-learning-based modeling attack.

The CRP data collection process generally involves feeding randomly generated challenges to the PUF instance, and collecting the responses. Though for some PUFs, such as Memristor-based PUFs, modeling attacks can be launched directly using challenges and responses. However, for most other PUF variants (e.g., for Arbiter PUFs), the machine-learning-based modeling attack fails when directly used with

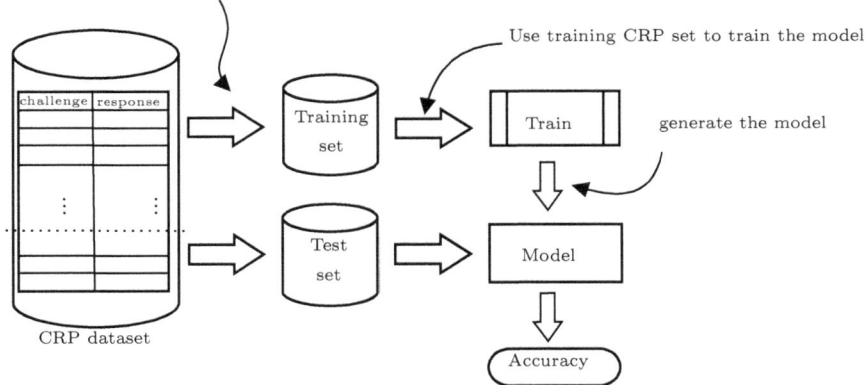

Fig. 4.1 Process of launching a modeling attack on PUF

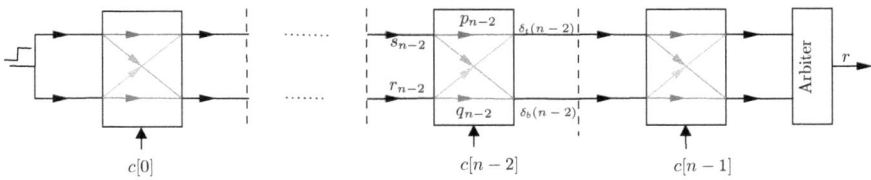

Fig. 4.2 Delay components in each stage of Arbiter PUF

challenges as input [3]. In such cases, it is important to identify the *features* that effectively capture the relationship between challenge and response. Next, we will discuss the mathematical model of the arbiter PUF, which will help us understand the features responsible for launching an attack.

4.2 Mathematical Model of Arbiter PUF (APUF)

It is popularly known as **linear additive delay model** of APUF, developed by D. Lim [1]. Let p_i, q_i, r_i, and s_i be four delay components of the ith delay stage in APUF, as shown in Fig. 4.2. Let $\delta_t(i)$ and $\delta_b(i)$ be the propagation delay of top signal from starting point to top path and bottom path at the end of ith delay stage, respectively.

For a given challenge $c \in \{+1, -1\}$, the recurrence relation of the propagation delay of the trigger signal at the output of the $(i+1)$th stage is given by

$$\delta_t(i+1) = \frac{1 + \mathbf{c}[i+1]}{2}(\delta_t(i) + p_{i+1}) + \frac{1 - \mathbf{c}[i+1]}{2}(\delta_b(i) + s_{i+1}) \quad (4.1)$$

4.2 Mathematical Model of Arbiter PUF (APUF)

$$\delta_b(i+1) = \frac{1+\mathbf{c}[i+1]}{2}(\delta_b(i) + q_{i+1}) + \frac{1-\mathbf{c}[i+1]}{2}(\delta_t(i) + r_{i+1}) \quad (4.2)$$

Now, we can estimate the delay difference ($\Delta(i+1)$) between top and bottom paths by subtracting Eq. (4.1) with Eq. (4.2), which is given by

$$\begin{aligned} \Delta(i+1) &= \delta_t(i+1) - \delta_b(i+1) \\ &= \frac{1+\mathbf{c}[i+1]}{2}(\Delta(i) + p_{i+1} - q_{i+1}) + \frac{1-\mathbf{c}[i+1]}{2}(\Delta(i) + r_{i+1} - s_{i+1}) \\ &= \Delta(i)\mathbf{c}[i+1] + \alpha_{i+1}\mathbf{c}[i+1] + \beta_{i+1} \end{aligned} \quad (4.3)$$

where

$$\alpha_{i+1} = \frac{p_{i+1} + r_{i+1} - q_{i+1} - s_{i+1}}{2}$$

$$\beta_{i+1} = \frac{p_{i+1} + s_{i+1} - q_{i+1} - r_{i+1}}{2}$$

Simplifying Eq. (4.3), we have

$$\begin{aligned} \Delta(-1) &= 0 \\ \Delta(0) &= \alpha_0 \mathbf{c}[0] + \beta_0 \\ \Delta(1) &= \Delta(0)\mathbf{c}[1] + \alpha_1 \mathbf{c}[1] + \beta_1 \\ &= (\alpha_0 \mathbf{c}[0] + \beta_0)\mathbf{c}[1] + \alpha_1 \mathbf{c}[1] + \beta_1 \\ &= \alpha_0 \mathbf{c}[0]\mathbf{c}[1] + \alpha_1 \mathbf{c}[1] + \beta_0 \mathbf{c}[1] + \beta_1 \\ \Delta(2) &= \Delta(1)\mathbf{c}[2] + \alpha_2 \mathbf{c}[2] + \beta_2 \\ &= (\alpha_0 \mathbf{c}[0]\mathbf{c}[1] + \alpha_1 \mathbf{c}[1] + \beta_0 \mathbf{c}[1] + \beta_1)\mathbf{c}[2] + \alpha_2 \mathbf{c}[2] + \beta_2 \\ &= \alpha_0 \mathbf{c}[0]\mathbf{c}[1]\mathbf{c}[2] + \alpha_1 \mathbf{c}[1]\mathbf{c}[2] + \alpha_2 \mathbf{c}[2] + \beta_0 \mathbf{c}[1]\mathbf{c}[2] + \beta_1 \mathbf{c}[2] + \beta_2 \\ &\cdots \\ \Delta(n-1) &= \Delta(n-2)\mathbf{c}[n-1] + \alpha_{n-1}\mathbf{c}[n-1] + \beta_{n-1} \\ &= \alpha_0 \mathbf{c}[0]\mathbf{c}[1]\mathbf{c}[2]\cdots\mathbf{c}[n-1] + \alpha_1 \mathbf{c}[1]\mathbf{c}[2]\cdots\mathbf{c}[n-1] \\ &\quad + \alpha_2 \mathbf{c}[2]\cdots\mathbf{c}[n-1] \\ &\quad + \cdots \\ &\quad + \alpha_{n-1}\mathbf{c}[n-1] + \beta_0 \mathbf{c}[1]\mathbf{c}[2]\cdots\mathbf{c}[n-1] + \beta_1 \mathbf{c}[2]\cdots\mathbf{c}[n-1] \\ &\quad + \cdots \\ &\quad + \beta_{n-1} \end{aligned} \quad (4.4)$$

Let the *parity vector* (Φ) defined by

$$\Phi[n] = 1 \text{ and } \Phi[i] = \prod_{j=i}^{n-1} \mathbf{c}[j] \tag{4.5}$$

Substituting Eq. (4.5) in Eq. (4.4), we get

$$\begin{aligned}\Delta(n-1) &= \alpha_0 \Phi[0] + (\alpha_1 + \beta_0)\Phi[1] + (\alpha_2 + \beta_1)\Phi[2] + \cdots \\ &\quad + (\alpha_{n-1} + \beta_{n-2})\Phi[n-1] + \beta_{n-1}\Phi[n] \\ &= \langle \Phi, \mathbf{w} \rangle.\end{aligned} \tag{4.6}$$

where $\mathbf{w} = (\alpha_0, \alpha_1 + \beta_0, \alpha_2 + \beta_1, \cdots, \alpha_{n-1} + \beta_{n-2}, \beta_{n-1})$.

Hence, the delay difference at the last stage ($\Delta(n-1)$) can be represented as the inner product of the parity vector corresponding to the applied challenge, and the constant vector $\mathbf{w}$. The corresponding response is given by

$$\begin{aligned}r &= \begin{cases} 1 & \text{if } \Delta(n-1) < 0, \\ 0 & \text{otherwise.} \end{cases} \\ &= sign(\Delta(n-1)) = sign(\langle \Phi, \mathbf{w} \rangle)\end{aligned} \tag{4.7}$$

If parity vector corresponding to each challenge can be considered as a data point in $n+1$-dimensional space, then w is the $n+1$-dimensional hyperplane that classifies the response. As a hyperplane can classify the responses of APUF instances, the responses are said to be linearly separable (has linear decision boundary). Hence, successful modeling of an APUF comes down to accurately predicting the hyperplane, i.e., weight vector $\mathbf{w}$. This can be achieved quite successfully by machine learning algorithms, viz., linear support vector machine (linear SVM) [1, 3] and logistic regression (LR) [3]. Please note that the feature vectors used for modeling attack are parity vectors and not challenge vectors.

4.3 Mathematical Model of XOR APUF

The XOR operation can be represented as product when inputs are bipolar encoded. Table 4.1 shows the relationship between bipolar multiplication and XOR operation.

For a given challenge $c \in \{+1, -1\}$, the response of an k-XOR APUF can be formulated using Eq. (4.7) as

$$F_{xor}(\mathbf{x}, k) = \text{sign}\left(\prod_{i=i}^{k} \mathbf{w_x}^T \cdot \Phi\right) \tag{4.8}$$

4.3 Mathematical Model of XOR APUF

Table 4.1 Relationship between bipolar multiplication and XOR

Boolean inputs		XOR	Bipolar encoded inputs		Multiplication
a	b	$a \oplus b$	a_e	b_e	$a_e \times b_e$
0	0	0	1	1	1
0	1	1	1	-1	-1
1	0	1	-1	1	-1
1	1	0	-1	-1	1

The modeling of F_{xor} thus involves computing a non-linear decision boundary in an $[(n+1)x]$-dimensional feature space [3]. The increase in k increases the non-linearity, making the learning difficult for k-XOR APUFs [3].

Based on Eq. (4.9), a machine learning model was constructed [3] using a sigmoid activation function and was trained using gradient descent algorithm. It is popularly referred to as the *logistic regression* (LR)-based modeling attack on XOR APUF. However, this model follows a very different structure from the linear logistic regression.

Initially, it was reported that modeling attack on x-XOR APUF is practically infeasible if $k \geq 6$ [3]. However, by employing substantial computational resources, the work [9] was able to attack for $k \leq 9$. It has been confirmed by experiment [9], where increasing the value of k results in exponential increase in training data and training time requirement. For example, the number of CRPs needed for launching a successful attack on an 9-XOR was 350 million, executed on a cluster with one TB of main memory.

4.3.1 Motivation of Applying Deep Learning to PUF Modeling Attack

Logistics Regression (LR) was also applied to model the XOR APUF, using the same mathematical XOR APUF model derived from that of the standalone APUF. When the ML model is defined in conjunction with a mathematical model of PUF, it is advantageous for the fact that the parameter values of the trained ML can be mapped easily to parameters of the PUF. For instance, using the LR attack model for XOR APUF, it is quite easy to determine the delay parameters for the constituent PUFs involved in XOR APUF. However, it is not always possible to construct a closed-form mathematical model for PUFs such as *chaotic PUF*. Also, ML modeling becomes progressively difficult as the underlying mathematical structure of the PUF variant becomes more complex. In order to overcome this problem, *Deep feedforward neural networks* (DFNN) offer a superior choice due to the following characteristics:

1. DFNN learns the input-output mapping without requiring a mathematical model.

2. DFNNs are capable of mapping highly non-linear complex input-output relationships and can easily be scaled by adding additional neurons and layers.

Next, we will discuss the DFNN-based modeling attacks.

4.4 DFNN Architecture for Modeling Attack on APUF Compositions

Santikellur et al. [3] were the first to use DFNN for modeling attacks on APUF compositions. Multiplexer PUF (MPUF) and Interpose PUF (iPUF), which were previously proven to be robust, were also broken using DFNN. There were subsequently several improvements proposed for each of the APUF variants that improved the existing DFNN attacks. A key aspect of the technique proposed in [3] is that it does not use special feature engineering for each APUF variant, but instead uses just "parity vectors" as input features and then uses DFNN's hierarchical representation capability to develop the automatic features for the modeling attack. Most importantly, the authors have made the dataset and machine learning software code available online.[1] Next, we will discuss the generic DFNN architecture, followed by the results of recently proposed DFNN-based modeling attacks on several APUF variants.

Figure 4.3 depicts the generic architecture of deep feedforward neural networks, used by authors [3] to launch modeling attacks for APUF compositions. In order to model the n-bit APUF variant, $(n+1)$-bit parity vector is fed as an input to the DFNN. This DFNN architecture consists of m hidden layers and one output bit for predicting a 1-bit response. Table 4.2 shows the hyperparameters used for the modeling attacks in [3].

Although DFNN are primarily used to launch attacks on robust non-linear PUFs, they can also be used on linearly separable PUF models, such as arbiter PUF. A key feature of DFNN is its ability to model PUF variants of varying robustness, something that makes it a convenient and comprehensive tool for carrying out modeling attacks. When compared with traditional ML techniques, DFNN typically requires more training CRPs for modeling. Next, we look at modeling attack results.

In Sect. 4.3, we established that arbiter PUF CRPs can be linearly separated if "parity vectors" extracted from challenges are employed as the feature. With parity vectors as input feature, arbiter PUF instance can be modeled using single output neuron without any hidden layer. In fact, a single output neuron with sigmoid activation function corresponds to linear logistic regression [8]. Table 4.3 shows the results of modeling attack on 64-bit APUF and 128-bit APUF.

Table 4.3 shows the modeling accuracy achieved and the time required to train the DFNN and LR model. The number of training CRPs is also presented. To model the 64-bit APUF using DFNN, 7000 training CRPs were required, while training

---

[1] https://github.com/Praneshss/Modeling_of_APUF_Compositions.

4.4 DFNN Architecture for Modeling Attack on APUF Compositions

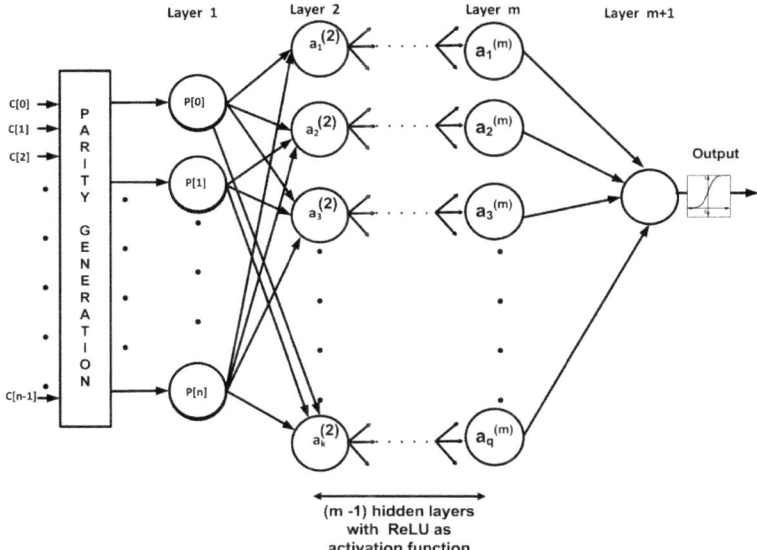

Fig. 4.3 Generic deep FF neural network architecture for modeling APUF compositions

Table 4.2 Hyperparameter values used [3]

Hyperparameters	Values
Kernel initializer	*Glurot uniform* [3]
Learning rate	*0.001*
Bias initializer	*Zeros*
Optimizer	*Adam* [3]
Loss function	*Binary cross entropy (BCE)*
Hidden layer activation function	*ReLU* [20]
Output layer activation function	*Sigmoid*

Table 4.3 Modeling accuracy result for APUF

Challenge Size (bits)	Attack Method	Training CRP Count	Prediction Accuracy (%)	Training Time (s)
64	DFNN	7,000	99.50	11.25
	LR-Rprop [3]	2,555	99%	0.13
128	DFNN	8,000	99.50	18.21
	LR-Rprop [3]	5,570	99%	0.51

the 128-bit APUF model required 8000 CRPs. In both 64-bit and 128-bit models, 99.5% accuracy was achieved. LR also achieves an accuracy of 99% for CRPs lower than those used by DFNN. As far as decision boundaries are concerned, both 64-bit and 128-bit are linearly separable, hence, there is no significant difference between training CRPs of 64-bit and 128-bit APUFs. Next, we will discuss the results of modeling attacks on composite PUFs.

4.4.1 Modeling Attack on XOR APUF

In Chap. 2, we explained that among several combination functions, XOR provides the best non-linearity when the PUFs are unbiased. There have been several modeling attacks that use LR and DFNN against simulated XOR APUFs. The authors of the work [3] carried out one of the first attacks on XOR APUF using LR. In [3], Rprop or RMSprop was employed as optimizer and it was reported that modeling attack on x-XOR APUF is infeasible if $x \geq 6$, although later careful implementation of LR was able to break x-XOR APUF for $x \leq 9$ [9], albeit employing substantial computational resources (e.g., a cluster with 1 TB of main memory), and substantial CRP data (e.g., 350 million CRPs for the 9-XOR APUF). An LR attack has recently been implemented in *Keras* using *Adam* optimization and *tanh activation* at the intermediate and final nodes. The modified LR is referred to as *LR-Adam*, while the LR attack implemented by [3] and [9] is referred to as *LR-Rprop*. The modified LR can be expressed as (compared to Eq. (4.9))

$$F_{xor}(\mathbf{x}, k) = \tanh\left(\prod_{i=i}^{k} \tanh(\mathbf{w_x}^T \cdot \Phi)\right) \tag{4.9}$$

Three works [3, 10, 11] have proposed DFNN-based attacks on XOR APUF. In contrast, the approach in [3] does not follow a specific DFNN structure as template, and it uses variable numbers of neurons and layers for attacking each XOR APUF variant. However, the [10] and [11] use three hidden layers of size $(2^k, 2^k, 2^k)$ and $(2^{k-1}, 2^k, 2^{k-1})$, respectively, for launching an attack on k-XOR PUF. Table 4.4 provides the hyperparameter settings used by the different authors for modeling XOR APUF.

The authors of [3] implemented the attack using Python 2.7 and the Keras 2.1.5 [19] framework, with TensorFlow [18] as the backend, and executed on a Linux workstation with 32 GB of main memory and a 3.3 GHz, 4-core Intel Xeon processor. All the experiments were conducted without us explicitly parallelizing the code across the cores, whereas the authors [12] used different setting using CPU/GPU with varying number of threads to implement DFNN-based attacks [10, 11].

Table 4.5 presents the modeling attack results of 64-bit and 128-bit XOR APUF. It should be noted that the results of works [10, 11] have been collected from [12], as their implementations were flawed [12].

4.4 DFNN Architecture for Modeling Attack on APUF Compositions

Table 4.4 Hyperparameters used for modeling XOR APUF

	Santikellur et al. DFNN [3]	Aseeri et al. DFNN [10]	Mursi et al. DFNN [11]	Rührmair et al. [3] Tobisch et al. [9] LR-Rprop	Wisiol et al. LR-Adam [12]
Architecture	varied arch.	$(2^k, 2^k, 2^k)$	$(2^{k-1}, 2^k, 2^{k-1})$	XOR APUF math. model	XOR APUF math. model
Hid. lay. activ.	ReLU	ReLU	Tanh	–	Tanh
Output activ.	Sigmoid	Sigmoid	Sigmoid	Sigmoid	tanh
Optimizer	Adam	Adam	Adam	RProp	Adam
Loss function	BCE	BCE	BCE	BCE	BCE
Learning rate	10^{-3}	10^{-3}	adaptive	RProp default	–
Initializer	Glorot Normal	Glorot Unif.	Gaussian	Gaussian	–

From Table 4.5, the following important aspects are to be noted:

1. With an increasing k value, the number of training CRP needed increases dramatically. LR attacks show a clear exponential dependence on training CRPs [9, 12].
2. Although the training CRP requirements for DFNN are higher than those for LR for lower k values, DFNN requires relatively fewer training CRPs for higher k values.

Table 4.5 can be used to compare the accuracy and training CRP dynamics of different attacks and different k values. However, comparison of the exact CRP count for different attacks might be misleading due to the following reasons:

1. Accuracy reported by different attacks has different values. This is important, since accuracy greater than 95% is considered a successful attack, the CRP requirements increase, sometimes dramatically [14], to achieve modeling accuracy of 98% to 99%.
2. The training time comparison across different implementations may be invalid because the implementations use different resources, but comparisons between several k-XOR APUFs can be made through the same work. Specifically for LR-Rprop, training time also increases significantly with k.
3. Most simulated APUF are configured with different values of additive random noise. These noise variations can impact learning accuracy, e.g., learning time is polynomial to inverse of noise [17].

The points outlined above are not only relevant for the XOR APUF but also for the comparison of other PUF modeling attacks. Aside from these points, it is important to point out that simulated PUFs are chosen for modeling attacks, primarily because implementation of PUFs with good performance characteristics on FPGA/ASIC is quite difficult. Any defective implementation will easily lead to easy modeling attack. Readers are advised to refer to detailed works explaining delay PUF's inconsistencies [16] and architectural bias [15].

Table 4.5 Modeling accuracy result for XOR APUF

Challenge Size (bits)	No. of XORs	Attack Method	Training CRP Count	Prediction Accuracy (%)	Training Time
64	2	DFNN* [3]	32×10^3	99.30	56.36 sec
	3	DFNN* [3]	36.8×10^3	99.22	1 min 12 sec
	4	DFNN* [3]	41.2×10^3	98.60	2 min 10 sec
		DFNN [10]	400×10^3	> 95	< 1 min
		DFNN [11]	150×10^3	> 95	< 1 min
		LR-Rprop [3]	12×10^3	99.00	3.42 min
		LR-Rprop* [9]	10×10^3	> 98	16 sec
		LR-Adam [12]	30×10^3	> 95	< 1 min
	5	DFNN* [3]	145×10^3	98.20	10 min 12 sec
		DFNN [10]	400×10^3	> 95	< 1 min
		DFNN [11]	200×10^3	> 95	< 1 min
		LR-Rprop [3]	80×10^3	99.00	2.08 hrs
		LR-Rprop* [9]	45×10^3	> 98	2.46 min
		LR-Adam [12]	260×10^3	> 95	4 min
	6	DFNN* [3]	680×10^3	97.68	20 min 52 sec
		DFNN [10]	2×10^6	> 95	< 1 min
		DFNN [11]	2×10^6	> 95	< 1 min
		LR-Rprop [3]	200×10^3	99.00	31.01 hrs
		LR-Rprop* [9]	210×10^3	> 98	30.34 min
		LR-Adam [12]	2×10^6	> 95	< 1 min
	7	DFNN [10]	5×10^6	> 95	< 1 min
		DFNN [11]	4×10^6	> 95	< 1 min
		LR-Rprop* [9]	3×10^6	> 98	2.43 hrs
		LR-Adam	20×10^6	> 95	3 min
	8	DFNN [10]	30×10^6	> 95	3 min
		DFNN [11]	6×10^6	> 95	13 min
		LR-Rprop* [9]	40×10^6	> 98	6.31 hrs
		LR-Adam [12]	150×10^6	> 95	28 min
	9	DFNN [10]	80×10^6	> 95	86 min
		DFNN [11]	45×10^6	> 95	16 min
		LR-Rprop* [9]	350×10^6	> 98	37.36 hrs
		LR-Adam [12]	500×10^6	> 95	14 min
	10	DFNN [11]	119×10^6	> 95	291 min
		LR-Adam [12]	1000×10^6	> 95	41 min
	11	DFNN [11]	325×10^6	> 95	1898 min
	2	DFNN* [3]	32×10^3	99.10	1 min 10 sec
	3	DFNN* [3]	37.6×10^3	98.90	2 min 5 sec

(continued)

4.4 DFNN Architecture for Modeling Attack on APUF Compositions   45

Table 4.5 (continued)

Challenge Size (bits)	No. of XORs	Attack Method	Training CRP Count	Prediction Accuracy (%)	Training Time
128	4	DFNN* [3]	255×10^3	97.80	8 min 30 sec
		DFNN [10]	400×10^3	> 95	< 1 min
		DFNN [11]	1000×10^3	> 95	< 1 min
		LR-Rprop [3]	24×10^3	99.00	2.53 hrs
		LR-Rprop* [9]	22×10^3	> 98	2.24 min
	5	DFNN [3]	655×10^3	97.87	29 min 21 sec
		DFNN [10]	3000×10^3	> 95	< 1 min
		DFNN [11]	1000×10^3	> 95	< 1 min
		LR-Rprop [3]	500×10^3	99.00	16.36 hrs
		LR-Rprop* [9]	325×10^3	> 98	12.11 min
	6	DFNN [10]	20×10^6	> 95	< 1 min
		DFNN [11]	10×10^6	> 95	< 1 min
		LR-Rprop* [9]	15×10^6	> 98	4.45 hrs
	7	DFNN [10]	40×10^6	> 95	5 min
		DFNN [11]	30×10^6	> 95	2 min
		LR-Rprop* [9]	400×10^6	> 98	66.53 hrs
	8	DFNN [10]	100×10^6	> 95	45 min

*reports the minimum number of training CRPs that were used to achieve a corresponding accuracy

4.4.2 Modeling Attack on Double Arbiter PUF (DAPUF)

DAPUF was initially proposed to improve the uniqueness and robustness of arbiter PUF [5]. DAPUF authors have demonstrated the robustness against SVM using less training CRPs. However, Yashiro et al. [13] and Khalafalla et al. [14] proposed two different DL-based attacks [13, 14] for modeling DAPUF. In [13], de-noising autoencoders are employed as intermediate layers in DFNN to evaluate the security of 64-bit 2-1, 3-1, and 4-1 DAPUF. The DFNN architecture consists of four layers and is trained in two phases: pre-training phase and fine-tuning phase. This deep learning architecture is trained in two phases: pre-training phase and fine-tuning phase. The first three layers use de-noising autoencoders to extract feature information during the pre-training phase using unsupervised learning, while the fourth layer uses logistic regression to perform fine-tuning for accurate classification. This attack was implemented using the PyLearn2 framework [21] and executed on a Linux workstation with 64 GB of main memory and a 2.67 GHz, 8-core Intel Xeon processor. On the other hand, the work [14] proposes a DFNN-based DL architecture which emulates DAPUF structure. For example, 3-1 DAPUF has six arbiter whose outputs are XOR-ed to produce the final single-bit response. Similarly, the authors [14] propose a six feedforward neural network with outputs connected to multiplication node. Each feedforward neural network consists of 12 fully connected layers each with 2000 neurons. Dropout is added at the end of each feedforward network to avoid overfitting. Figure 4.4 shows the DFNN-based architecture for modeling 3-1 DAPUF. For 4-1 DAPUF, 12 feedforward network are used, where each corresponds to arbiter output of 4-1 DAPUF. However, authors used 18 fully connected layers of 2000 neurons for each feedforward network. This attack was executed on Nvidia GeForce GTX 1080 Ti GPU and 11GB RAM. The authors [14] also implemented LR-based attack following linear LR model for 2-1 DAPUF and following XOR math model for 3-1 DAPUF and executed on Intel 8th Gen I7-8250 CPU with 16GB RAM of memory.

Table 4.6 shows the results of modeling attacks on various 64-bit DAPUF. From the table, we can notice that even with 17M CRPs, the accuracies achieved for 3-1 DAPUF and 4-1 DAPUF are 86% and 81.5%.

4.4.3 Modeling Attack on Multiplexer PUF and Its Variants

MPUF and its variants are proposed as alternatives to XOR APUF with increased reliability and greater robustness. The authors of MPUF [6] concluded through extensive theoretical and experimental analysis that the (64, 3)-rMPUF to have comparable robustness to modeling attack as an 10-XOR APUF with 64-bit challenge, and its reliability is as high as the reliability of a 4-XOR APUF. The authors demonstrate that MPUF variants can also achieve statistical properties similar to LSPUF without using any additional input network. Santikellur et al. [3] proposed novel modeling attack on MPUF and its variants, i.e., cMPUF and rMPUF (for the design parameter $k = 3, 4, 5$). It was found that these variants can be modeled with relatively lesser number of CRPs than XOR-based APUF compositions. Table 4.7 shows the employed DFNN architecture and Table 4.8 shows the modeling attack results of MPUF and its variants for both 64-bit and 128-bit challenge sizes. From the table, it is evident that the complexity of the modeling attack increases as the value of k increases.

In contrast to the conclusions reached in [6], robustness to modeling attacks of rMPUFs seems inferior to that of XOR APUFs.

4.4.4 Modeling Attack on Interpose PUF (iPUF)

As described in Sect. 2.3.4 earlier, Interpose PUF was recently proposed robust PUF which uses XOR APUF as constituent PUF. The authors [7] claimed through theoretical analysis and experimental results that an (x, y)-iPUF design while having the same hardware overhead and comparable reliability as an $(x + y)$-XOR PUF provides much higher robustness to modeling attacks. Earlier, the authors claimed that using the middle bit of the second APUF as the interpose position, the IPUF is robust against classical-machine-learning-based modeling attack, reliability-based

4.4 DFNN Architecture for Modeling Attack on APUF Compositions

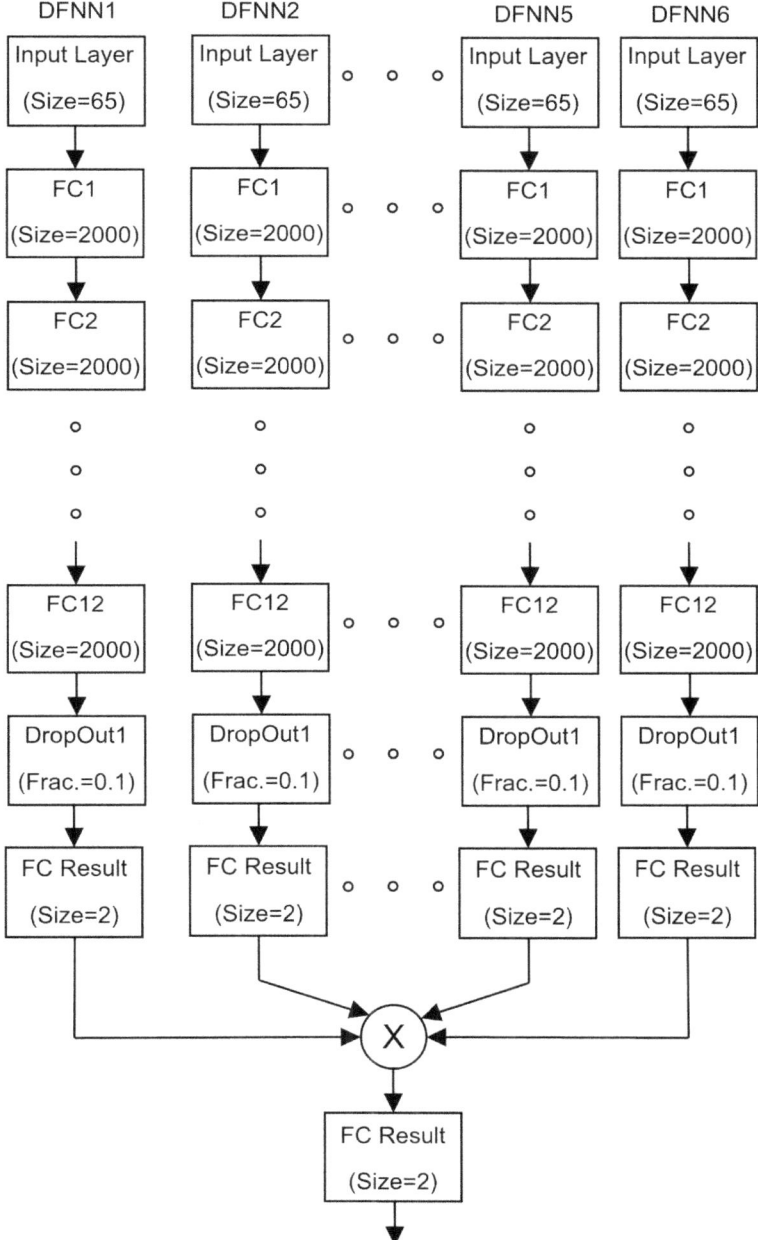

Fig. 4.4 DFNN-based architecture for modeling 64-bit 3-1 DAPUF

Table 4.6 Modeling accuracy result for 64-bit DAPUF

DAPUF	Attack Method	Training CRP Count in 10^3	Prediction Accuracy (%)	Training Time
2-1	LR [14]	100×10^3	93.4	–
	DFNN [13]	40×10^3	90	–
3-1	DFNN [14]	4×10^6	76.2	–
	DFNN [13]	50×10^3	68	2 hr 25 min
	DFNN [14]	17×10^6	86	approx. 5 hrs
4-1	DFNN [13]	50×10^3	63	2 hr 25 min
	DFNN [14]	17×10^6	81.5	–

Table 4.7 DFNN architecture for MPUF and variants for different k values

k values	PUF Type	Challenge Size (bits)	No. of Hidden layers	Av. No. of Nodes per hidden layer
3	MPUF	64	2	30
		128	4	27
	cMPUF	64	4	27
		128	4	27
	rMPUF	64	3	30
		128	4	30
4	MPUF	64	3	50
		128	3	50
	cMPUF	64	4	27
		128	4	40
	rMPUF	64	3	50
		128	4	50
5	MPUF	64	3	55
		128	4	65
	cMPUF	64	4	31
		128	3	60
	rMPUF	64	3	70
		128	4	65

modeling attack, and cryptanalytic attacks. In particular, the authors demonstrate that the (3, 3)-iPUF is satisfactorily resistant to all known attacks.

Santikellur et al. [3] were the first to propose DFNN-based attacks against the Interpose PUF, showing that various iPUFs up to (4, 4)-iPUF are vulnerable to DFNN attacks. Later, various works have extended these attacks to higher values of iPUF. Later, the original proposers of iPUF in [7] claimed that (k_{up}, k_{down})-iPUFs

4.4 DFNN Architecture for Modeling Attack on APUF Compositions

Table 4.8 Modeling accuracy result for MPUF and variants for different k values

k value	PUF Type	Challenge Size (bits)	Training CRP Count in 10^3	Prediction Accuracy (%)	Training Time (s)
3	MPUF	64	111	98.10	2 min 5
		128	112	97.50	3 min 23
	cMPUF	64	112	98.30	5 min 37
		128	112	97.50	4 min 5
	rMPUF	64	80	98.20	5 min 3
		128	80	97.40	5 min 40
4	MPUF	64	176	97.44	4 min 31
		128	184	96.49	16 min 10
	cMPUF	64	112	97.36	5 min 07
		128	160	97.14	8 min 25
	rMPUF	64	184	97.12	9 min 31
		128	264	96.23	20 min 16
5	MPUF	64	256	97.02	14 min 13
		128	312	96.40	22 min 43
	cMPUF	64	152	97.24	10 min 21
		128	215	96.36	10 min 13
	rMPUF	64	320	96.54	15 min 23
		128	400	95.45	32 min 27

are comparably secure to an $(k_{down} + \frac{k_{up}}{2})$-XOR APUF. Primarily, these modeling attacks on iPUF are mainly LR based and DFNN based.

Direct XOR-based LR modeling attacks cannot be launched since the interpose bit information is absent from the training CRP dataset. Recently, a novel modeling attack strategy based on LR has been proposed, called *Splitting attack* [2], in which attacks can be built on upper layer XOR APUF and lower layer XOR APUF separately. The attack strategy can be divided into three steps:

1. Modeling of lower layer XOR APUF using linearization attack: To model the lower XOR APUF, first, a new CRP dataset is created by replacing interpose bit with random bit. This technique of bypassing interpose bits is called linearization attack. Furthermore, LR-based modeling is employed with the new dataset to achieve maximum accuracy of 75.0%.
2. Modeling of upper layer XOR APUF: A new dataset can be heuristically constructed for upper layer XOR APUF using the CRP dataset of complete iPUF and generated dataset at step 1 (provided decent accuracy is achieved in step 1). Again, LR is applied to this newly generated dataset for modeling upper layer XOR APUF.
3. By utilizing independent ML models of upper layer and lower layer XOR APUF, the accuracy of the PUF is improved, i.e., with the constructed upper layer XOR APUF model (step 2), an improved lower layer dataset is generated. Based on the

Algorithm 1: Pseudo-code for Generalized Splitting attack on iPUF
Input : iPUF parameters (n, k_{up}, k_{down}, i) and CRP dataset (C, r) where $C = (c_1, \ldots, c_i, c_{i+1}, \ldots, c_n)$ **Output::** iPUF model   **procedure** SplittingAttack$(n, k_{up}, k_{down}, i, C, r)$

1   Generate the training dataset (C_{down}, r) for lower layer XOR APUF k_{down} by replacing interpose bit with random chosen bit $\mathbf{c} \xleftarrow{R} \{0, 1\}$.
       $C_{down} = \{(c_1, \ldots, c_i, \mathbf{c}, c_{i+1}, \ldots, c_n)\}$
 2   Train the ML model f_{down} for lower layer XOR APUF using the dataset (C_{down}, r).
       while *test_accuracy* < 95% **do**
 3   Generate upper layer training dataset (C, r_{up}) where r_{up} is heuristically selected using f_{down} and (C_{down}, r).
 4   Train the ML model f_{up} for upper layer XOR APUF using the dataset (C, r_{up}).
 5   Generate the training dataset (C_{down}, r) for lower layer XOR APUF f_{down} using upper layer XOR APUF model f_{up}.
 6   Train the ML model f_{down} for lower layer XOR APUF using the dataset (C_{down}, r).
       end
       return iPUF model $f_{down}(c_1, \ldots, c_i, f_{up}(C), c_{i+1}, \ldots, c_n)$

improved lower layer dataset, a more accurate lower layer XOR APUF model is constructed that again helps in the generation of better upper layer dataset. The process stops only when the expected test accuracy has been achieved.

The authors empirically show that using splitting attack, iPUF is (k_{up}, k_{down})-iPUF which is comparably secure to an max$\{k_{down}, k_{up}\}$-XOR APUF. Although splitting was first performed with LR, it was later found that using DFNN as a drop-in replacement for LR increased the attack's efficiency [12, 16]. Algorithm 1 shows the pseudo-code for the generalized splitting attack on iPUF.

Table 4.9 shows the modeling attack results on iPUF. It includes the results of modeling attacks carried out on real datasets and simulated datasets. The DFNN-Splitting and LR-Splitting attack methods refer to splitting attacks implemented with DFNN and LR, respectively, and DFNN represents the traditional modeling attack that performs modeling attack on iPUF without utilizing the knowledge of its structure (a sort of black-box modeling). With black-box DFNN, up to 64-bit (4, 4)-iPUF was broken with 320K CRPs. Larger iPUFs such as (1, 11) and (11, 11)-iPUFs were successfully attacked using DFNN-splitting method with 350M and 650M CRPs. Even though LR-Splitting proved to be able to break (1, 9)-iPUF with 750 CRPs, the authors [16] reported that LR-Splitting performed poorly on real datasets, especially when PUF is biased. As can be seen, black-box DFNN outperforms LR-Splitting on the (1, 5)-iPUF real dataset and DFNN-Splitting enhances the attack with both improved accuracy and less training time compared to the other two. It is also clear that DFNN-Splitting attack is significantly more effective for simulated iPUFs where LR-Splitting performs modeling attack on (1, 5)-iPUF with 20 million CRPs but DFNN-Splitting successfully breaks (1, 5)-iPUF with 6 million CRPs.

DFNN on simulated dataset [3] used three hidden layers with 50 and 60 neurons per layer for modeling (3, 3)-iPUF and (4, 4)-iPUF, respectively, and DFNN on

4.4 DFNN Architecture for Modeling Attack on APUF Compositions

Table 4.9 Modeling accuracy result for IPUF

CRP dataset Source	Attack Method	Challenge Size (bits)	IPUF type (x, y)	Training CRP Count	Prediction Accuracy (%)	Training Time
Simulation [3]	DFNN	64	(3, 3)	240×10^3	98.30	6 min 29 sec
			(4, 4)	319×10^3	97.44	5 min 23 sec
		128	(3, 3)	288×10^3	97.47	10 min 21 sec
			(4, 4)	647×10^3	97.68	32 min 17 sec
Simulation [2]	LR-Splitting	64	(1, 5)	0.5×10^6	> 95	10 min 22 sec
			(1, 6)	2×10^6	> 95	1 hr 29 min
			(1, 7)	20×10^6	> 95	20 hr 5 min
			(1, 9)	750×10^6	> 95	approx 8 weeks
			(5, 5)	1×10^6	> 95	14 min 36 sec
			(6, 6)	5×10^6	> 95	2 hr 13 min
			(7, 7)	40×10^6	> 95	17 hr 131 min
			(8, 8)	150×10^6	> 95	11 days
Simulation [12]	DFNN-Splitting	64	(1, 7)	6×10^6	> 95	< 1 hr
			(1, 11)	350×10^6	> 95	–
			(11, 11)	650×10^6	> 95	–
FPGA [16]	LR-Splitting	64	(1, 5)	1×10^6	71.07	11 min
	DFNN		(1, 5)	1×10^6	87.54	42 min
	DFNN-Splitting		(1, 5)	1×10^6	93.06	15 min

real dataset [16] used three hidden layers with 64 neurons per layer for modeling (1, 5)-iPUF. In a DFNN-Splitting attack on simulated dataset, each k-XOR APUF was modeled using three hidden layers with $\left(2^{k-1}, 2^k, 2^{k-1}\right)$ neurons. Please note that modeling attack on silicon dataset of iPUF was carried out on machine with 96 CPU cores with maximum clock frequency of 2.3 GHz and 256 GB of memory.

References

1. Lim, D. (2004). *Extracting secret keys from integrated circuits*. Master's thesis, Massachusetts Institute of Technology, U.S.A.
2. Wisiol, N., Mühl, C., Pirnay, N., Nguyen, P. H., Margraf, M., Seifert, J. -P., van Dijk, M., & Rührmair, U. (2019). Splitting the interpose PUF: A novel modeling attack strategy. *Cryptology ePrint Archive*, Report 2019/1473, https://eprint.iacr.org/2019/1473.
3. Rührmair, U., Sehnke, F., Sölter, J., Dror, G., Devadas, S., & Schmidhuber, J. (2010). Modeling attacks on physical unclonable functions. In *Proceedings of the ACM Conference on Computer and Communications Security, Ser. CCS'10*, 237–249.
4. Majzoobi, M., Koushanfar, F., & Potkonjak, M. (2008). Lightweight secure PUFs. In *Proceedings of the IEEE/ACM International Conference on Computer-Aided Design, Ser. ICCAD'08*, 670–673.
5. Machida, T., Yamamoto, D., Iwamoto, M., & Sakiyama, K. (2015). A new arbiter PUF for enhancing unpredictability on FPGA. *The Scientific World Journal, 2015*.
6. Sahoo, D. P., Mukhopadhyay, D., Chakraborty, R. S., & Nguyen, P. H. (2018). A multiplexer-based arbiter PUF composition with enhanced reliability and security. *IEEE Transactions on Computers, 67*(3), 403–417.
7. Nguyen, P. H., Sahoo, D. P., Jin, C., Mahmood, K., Rührmair, U., & van Dijk, M. (2018). *The interpose PUF: Secure PUF design against state-of-the-art machine learning attacks*. Accessed May 2018, from https://eprint.iacr.org/2018/350.
8. Friedman, J., Hastie, T., Tibshirani, R., et al. (2001). The elements of statistical learning. *Springer Series in Statistics New York, 1*(10).
9. Tobisch, J., & Becker, G. T. (2015). On the scaling of machine learning attacks on PUFs with application to noise bifurcation. In S. Mangard & P. Schaumont (Eds.), *Proceedings of International Workshop on Radio Frequency Identification: Security and Privacy Issues, Ser. RFIDSec'15* (pp. 17–31).
10. Aseeri, A. O., Zhuang, Y., & Alkatheiri, M. S. (2018). A machine learning-based security vulnerability study on XOR PUFs for resource-constraint internet of things. *In IEEE International Congress on Internet of Things, 2018*, 49–56.
11. Mursi, K. T., Thapaliya, B., Zhuang, Y., Aseeri, A. O., & Alkatheiri, M. S. (2020). A fast deep learning method for security vulnerability study of XOR PUFs. *Electronics, 9*(10), 1715.
12. Wisiol, N., Mursi, K. T., Seifert, J.-P., & Zhuang, Y. (2021). Neural-network-based modeling attacks on XOR arbiter PUFs revisited. *IACR Cryptology Eprint Archive, 2021*, 555.
13. Yashiro, R., Machida, T., Iwamoto, M., & Sakiyama, K. (2016). Deep-learning-based security evaluation on authentication systems using arbiter puf and its variants. In K. Ogawa & K. Yoshioka (Eds.), *Advances in Information and Computer Security* (pp. 267–285). Springer International Publishing.
14. Khalafalla, M., & Gebotys, C. (2019). PUFs deep attacks: Enhanced modeling attacks using deep learning techniques to break the security of double arbiter PUFs. In *Design, Automation & Test in Europe Conference & Exhibition (DATE)* (pp. 204–209).
15. Sahoo, D. P., Nguyen, P. H., Chakraborty, R. S., & Mukhopaday, D. (2016). On the architectural analysis of arbiter delay PUF variants. *Cryptology ePrint Archive, Report 2016/057*, https://eprint.iacr.org/2016/057.

References

16. Aghaie, A., & Moradi, A. (2021). Inconsistency of simulation and practice in delay-based strong PUFs. *IACR Cryptology ePrint Archive, 2021*, 482.
17. Yu, M.-D., M'Raïhi, D., Verbauwhede, I., & Devadas, S. (2014). A noise bifurcation architecture for linear additive physical functions. *In IEEE International Symposium on Hardware-Oriented Security and Trust (HOST), 2014*, 124–129.
18. Abadi, M., et al. (2016). TensorFlow: A system for large-scale machine learning. In *Proceedings of the USENIX Conference on Operating Systems Design and Implementation, Ser. OSDI'16* (pp. 265–283).
19. Chollet, F., et al. (2015). Keras: The python deep learning library, https://keras.io.
20. Nair, V., & Hinton, G. E. (2010). Rectified linear units improve restricted boltzmann machines. In *Proceedings of the International Conference on Machine Learning, Ser. ICML'10* (pp. 807–814).
21. Goodfellow, I. J., Warde-Farley, D., Lamblin, P., Dumoulin, V., Mirza, M., Pascanu, R., Bergstra, J., Bastien, F., & Bengio, Y. (2013). Pylearn2: A machine learning research library. http://arxiv.org/abs/1308.4214.

Chapter 5
Improved Modeling Attack on PUFs based on Tensor Regression Network

5.1 Introduction

In the previous chapter, we have discussed DFNN-based modeling attacks on various APUF compositions. The purpose of this chapter is to understand design and development of an improved machine learning model for launching modeling attacks. While DFNN-based attacks provide an effective method of launching an attack, they are often regarded as "black-box" models due to the difficulty of interpreting them. Instead, it is a good option to build ML models that take into account the structure of the dataset. Indeed, as opined in [8], "It is not the data that should fit models, but models that should fit the data". In this chapter, we study one such design of a customized ML model to launch modeling attack on PUF variants which have traditionally proved robust against such attacks. We develop the attack based on the mathematical theory of XOR APUF, and later extend it to attack some other related PUF variants, e.g., LSPUF.

XOR APUFs have been shown to be robust to machine-learning-based modeling and the robustness increases significantly as the number of APUFs being XOR-ed increases. The Logistic-Regression-based modeling of XOR APUFs has a severe limitation, because of its exponential dependence on the number of training CRPs required [16] and also its exponential dependence on the number of CRPs. In this chapter, we discuss a novel customized model that aims to develop a computationally efficient way to model XOR APUF. It interprets the problem of XOR APUF modeling from a tensor computation point of view. In Chap. 3, we discussed the mathematical model of the XOR APUF (ref. Eq. (4.9)). It can also be reformulated in terms of tensor products, as noted in [2]:

$$f_{xor} = \text{sign}\left(\bigotimes_{i=1}^{x} \mathbf{w_x} \ldots \bigotimes_{i=1}^{x} \Phi\right) = \text{sign}\left(w_{xor} \ldots \Phi_{xor}\right) \qquad (5.1)$$

where $\Phi_{xor} = \bigotimes_{i=1}^{x} \Phi$ and $w_{xor} = \bigotimes_{i=1}^{x} \mathbf{w_x}$, and "$\bigotimes$" denotes the tensor product operator. This performs a linear classification of XOR APUF responses using a separating hyperplane in an $(n+1)^x$-dimensional feature space.

Our improved model is based on the tensor product interpretation as presented in Eq. (5.1), and is termed "Efficient CP-decomposition-based Tensor Regression Network (ECP-TRN)". It incorporates two important tensor concepts: Tensor Regression Network (TRN) and CANDECOMP/PARAFACT Tensor Decomposition (CP-decomposition). As a pre-requisite to understanding these concepts, we will revisit the basics of tensors and related topics in the following section.

5.2 Tensor Basics

Below are several important concepts related to tensors and TRN, as well as CP-decomposition. They have been adapted mostly from [11].

- TENSOR: Tensors are easy to understand when we see them as generalizations of scalars, vectors, and matrices. As an example, a scalar represents an order-0 tensor, a vector an order-1 tensor, a matrix an order-2 tensor, etc. In this sense, an order-N tensor is a mathematical structure that can be represented by a multidimensional array. An order-N tensor $\mathcal{X} \in \mathbb{R}^{I_1 \times I_2 \times \cdots \times I_N}$ comprises each element denoted by $x_{i_1, i_2, \ldots, i_N}$ and for $n = 1, 2, \ldots, N$, the dimension of the tensor $\mathcal{X}$ along the nth node is I_n.
- TENSOR PRODUCT: The product of two tensors $\mathcal{X} \in \mathbb{R}^{I_1 \times I_2 \times \cdots \times I_N}$ and $\mathcal{Y} \in \mathbb{R}^{I'_1 \times I'_2 \times \cdots \times I'_M}$ is defined as another tensor $(\mathcal{X} \otimes \mathcal{Y})_{i_1, i_2, \ldots, i_N, i'_1, i'_2, \ldots, i'_M} = x_{i_1, i_2, \ldots, i_N} y_{i'_1, i'_2, \ldots, i'_M}$, for all index values.
- INNER PRODUCT OF TENSORS: The inner product of two same-sized tensors is a generalization of vectors where $\mathcal{X}, \mathcal{Y} \in \mathbb{R}^{I_1 \times I_2 \times \cdots \times I_N}$ denoted by $\langle \mathcal{X}, \mathcal{Y} \rangle$ is sum of the product of corresponding entries:

$$\langle \mathcal{X}, \mathcal{Y} \rangle = \sum_{i_1=1}^{I_1} \sum_{i_2=1}^{I_2} \cdots \sum_{i_N=1}^{I_N} x_{i_1, i_2, \ldots, i_N} y_{i_1, i_2, \ldots, i_N} \qquad (5.2)$$

- TENSOR RANK: The notion of the rank of a tensor is similar to that of the rank of a matrix in linear algebra. Accordingly, the "rank-one tensors" can be considered to be tensors which can be represented as outer products of same-sized vectors and rank of a tensor $\mathcal{X}$ can be defined as the smallest number of rank-one tensors that fits tensor $\mathcal{X}$ exactly.

5.2 Tensor Basics 57

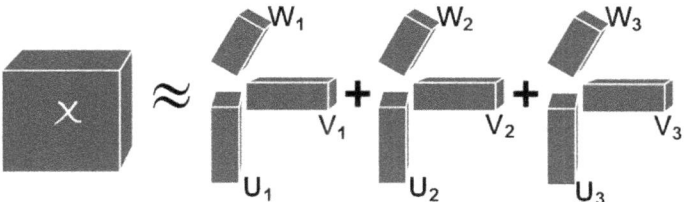

Fig. 5.1 CP-decomposition of an order-3 tensor. It is approximated by the sum of rank-one tensors. The rank of a tensor here is 3

5.2.1 CP-Decomposition

The use of low-rank tensors to express a tensor has been a huge interest since it reveals underlying structures in the tensor. Tensor decomposition techniques are used to express a tensor $\mathcal{X}$ using low-rank tensors. Tensor decompositions are easier to understand when viewed as generalizations of matrix decomposition. You might be wondering if the popular algorithm of matrix decomposition known as *Singular Value Decomposition* (SVD) can be applied to tensors. In literature, there are two important types of tensor decomposition that are derived from generalized properties of matrix SVD: The *Tucker decomposition* and the *CANDECOMP/PARAFAC decomposition* (commonly known as CP-decomposition). We will discuss the latter one since it is more relevant and provides a foundation for understanding ECP-TRN.

- CP-DECOMPOSITION is a higher order generalization of matrix SVD. CP-decomposition algorithm decomposes a tensor $\mathcal{X}$ into a sum of rank-one tensors. If, for instance, $\mathcal{X}$ is an order-3 tensor and R is the tensor rank, then

$$\mathcal{X} \approx \sum_{r=1}^{R} \mathbf{U}_r \otimes \mathbf{V}_r \otimes \mathbf{W}_r \qquad (5.3)$$

Figure 5.1 illustrates the CP-decomposition of the order-3 tensor. **A CP-decomposed representation needs a much smaller amount of storage than full tensor. To illustrate, an order-3 tensor** $\mathcal{X}$ **with dimensions** $I \times I \times I$ **requires** I^3 **storage, whereas a CP-decomposed representation requires** $I \ldots 3 \ldots R$ **amount of storage.**

- UPPER BOUND FOR CP TENSOR RANK: The work [12] provides upper bound for the CP tensor rank R of a tensor $\mathcal{X} \in \mathbb{R}^{I_1 \times I_2 \times \cdots \times I_N}$ and it is is given by

$$R \leq \left(\prod_{\ell=1}^{N} I_\ell \right) / \max_{\ell} I_\ell$$

5.2.2 Tensor Regression Networks

A *Tensor Regression Network* (TRN) is a specialized kind of neural network architecture that directly uses tensors as trainable components. They were introduced by the work [18] for solving the image classification problem. The work [18] proposes two trainable components: The TCL (tensor contraction layer) and the TRL (tensor regression layer). The architecture consists of two types of layers: *Tensor Contraction Layer* (TCL) and *Tensor Regression Layer* (TRL). The main objective of this work is to exploit the multi-linear structure in CNN for image datasets without sacrificing the efficiency of neural networks. In this work, fully connected layers are replaced after the convolution layers in CNN with TCLs and TRLs, in order to preserve the data's multi-modal structure. Additionally, it tries to reduce the number of parameters by enforcing a low tensor rank structure without sacrificing much accuracy. Various works have used different tensor decomposition methods to enforce weight tensors in low-rank space. Specifically, in the work [18], Tucker decomposition is applied in TRNs, while the work [14] analyzes the effectiveness of various decomposition methods, such as CP, Tucker, and Tensor Train (TT) decomposition methods. According to their analysis, the work [14] reports that CP-decomposition can be extremely useful for getting greater compression for image datasets while compromising small reductions in accuracy.

It is now possible to describe the ECP-TRN approach in more detail.

5.3 ECP-TRN: Efficient CP-Decomposition-Based Tensor Regression Networks

As with linear regression that involves computing the weight vector to determine the optimal hyperplane, the tensor regression computes the weight tensor $\mathcal{W}$ and the scalar $b \in \mathbb{R}$ to determine the optimal separating hyperplane. It is typically expressed as

$$f(\mathcal{X}) = \langle \mathcal{W}, \mathcal{X} \rangle + b \tag{5.4}$$

which is easily recognizable to be a generalization of determining the w_{xor} tensor in Eq. (5.1) (which implicitly has $b = 0$) to solve the XOR APUF modeling problem.

When dealing with high-dimensional tensors, not only do computations become more difficult, but also space consumption becomes an issue. We note that the following factors facilitate computationally efficient XOR APUF modeling according to Eq. (5.1):

- Using a tensor decomposition technique such as CP-decomposition, we can approximate the tensor $\mathcal{W}$ as the sum of rank-one tensors, where each rank-one tensor is an outer product of vectors. This approximate version of $\mathcal{W}$ usually has a rank that is much smaller than the theoretical upper bound of $\mathcal{W}$, as supported by our experimental findings (see Sect. 5.4).

5.3 ECP-TRN: Efficient CP-Decomposition-Based Tensor Regression Networks

- The input $\mathcal{X} = \Phi_{xor}$ is by definition a rank-one tensor since it can be represented using the outer product of vectors.

These alternative representations result in computations involving $\mathcal{W}$ and $\mathcal{X}$ having a low space complexity. In the case where $\mathcal{W}$ is expressed as a sum of rank-one tensors, we can simplify the computation of the inner product $\langle \mathcal{X}, \mathcal{W} \rangle$ without even having to reconstruct each of them from the decomposed rank-one tensors. Through the synergy of these two techniques, efficiency in computation is achieved, and it is termed as "Efficient CP-decomposition-based Tensor Regression Network" (ECP-TRN). Here is a more detailed explanation of the above claims.

An x-XOR APUF with n-bit challenges and m-bit output has the following parity tensor ($\mathcal{X}$):

$$\mathcal{X} = \Phi_{xor} \in \mathbb{R}^{\underbrace{(n+1) \times (n+1) \times \cdots \times (n+1)}_{x \text{ times}}} \tag{5.5}$$

and regression weight tensor ($\mathcal{W}$):

$$\mathcal{W} = w_{xor} \in \mathbb{R}^{\underbrace{(n+1) \times (n+1) \times \cdots \times (n+1)}_{x \text{ times}} \times 1} \tag{5.6}$$

where the last dimension $m = 1$ is considered which indicates that an XOR APUF has one response bit and the last dimension represents the fact that an XOR APUF has $m = 1$ response bits.

When $\mathcal{W}$ is represented as approximation as a sum of rank-one tensors of the form $\mathbf{w}_{r1} \otimes \mathbf{w}_{r2} \otimes \cdots \otimes \mathbf{w}_{rx} \otimes \mathbf{O}_r$, it is represented as

$$\mathcal{W} \approx \sum_{r=1}^{R} \mathbf{w}_{r1} \otimes \mathbf{w}_{r2} \otimes \cdots \otimes \mathbf{w}_{rx} \otimes \mathbf{O}_r \tag{5.7}$$

and $\mathcal{X}$ as

$$\mathcal{X} = \Phi_{xor} = \bigotimes_{i=1}^{x} \Phi \tag{5.8}$$

In this model, R is a hyperparameter that must be tuned because calculating the exact tensor rank is known to be *NP-complete problem* problem [15]. A well-tuned rank R will, however, have a low approximation error (ϵ) resulting from the decomposition in Eq. (5.7), facilitating model convergence.

It would be simpler yet to express the computation of the inner product $\langle \mathcal{W}, \mathcal{X} \rangle$ in terms of the rank-one tensors instead of reconstructing the (approximate) $\mathcal{X}$ and $\mathcal{W}$ tensors. The same problem was addressed in [16], and a solution is provided below. Let tensors $\mathcal{A}$ and $\mathcal{B}$ be decomposed based on the CP-decomposed form and expressed as

$$\mathcal{A} = \sum_{r=1}^{R_1} \mathbf{a}_{r1} \otimes \mathbf{a}_{r2} \otimes \cdots \otimes \mathbf{a}_{rN} \tag{5.9}$$

$$\mathcal{B} = \sum_{r=1}^{R_2} \mathbf{b}_{r1} \otimes \mathbf{b}_{r2} \otimes \cdots \otimes \mathbf{b}_{rN} \tag{5.10}$$

respectively, where rank of $\mathcal{A}$ is R_1 and rank of $\mathcal{B}$ is R_2, then:

$$\langle A, B \rangle = \sum_{p=1}^{R1} \sum_{q=1}^{R2} \langle \mathbf{a}_{p1}, \mathbf{b}_{q1} \rangle \langle \mathbf{a}_{p2}, \mathbf{b}_{q2} \rangle \cdots \langle \mathbf{a}_{pN}, \mathbf{b}_{qN} \rangle \tag{5.11}$$

Substituting Eq. (5.11) into (5.4), we get

$$f(\mathcal{X}) = \sum_{r=1}^{R} \langle \Phi, \mathbf{w}_{r1} \rangle \langle \Phi, \mathbf{w}_{r2} \rangle \cdots \langle \Phi, \mathbf{w}_{rx} \rangle \langle \Phi, \mathbf{O}_r \rangle + b \tag{5.12}$$

Connecting (5.12) with Eq. 5.1, a succinct formulation of the problem of modeling an x-XOR APUF is: for given components of the parity vector Φ, learn the vectors $\mathbf{w}_{r1}$, $\mathbf{w}_{r2}$, ..., $\mathbf{w}_{rx}$, $\mathbf{O}_r$ by ECP-TRN, so that the prediction of the x-XOR PUF response can be carried out using the equation:

$$\begin{aligned} f_{xor} &= f(\Phi_{xor}) \\ &= \text{sign}\left(\sum_{r=1}^{R} \langle \Phi, \mathbf{w}_{r1} \rangle \langle \Phi, \mathbf{w}_{r2} \rangle \cdots \langle \Phi, \mathbf{w}_{rx} \rangle \langle \Phi, \mathbf{O}_r \rangle \right) \end{aligned} \tag{5.13}$$

A figurative illustration of using ECP-TRN to model 3-XOR APUF is shown in Fig. 5.2.

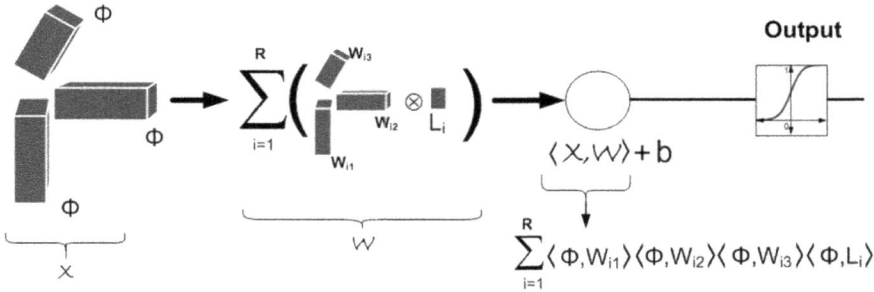

Fig. 5.2 ECP-TRN model for 3-XOR APUF

Table 5.1 ECP-TRN model characteristics

Characteristics	Values
Loss function	*Binary cross entropy*
Weight initializer	*Xavier normal* [1]
Regularization	*L2*
Bias initializer	*Zeros*
Optimizer	*Adam* [2]
Output layer activation function	*Sigmoid*

5.4 Experimental Results

5.4.1 Simulation and Modeling Setup

Experimental results are reported for CRPs collected from MATLAB simulation models of the PUFs. In line with the paper [2, 5], all delay parameters are positive, independent, and identically distributed, and that they follow a normal distribution with mean $\mu = 0.1$ and $\sigma = 1$. In addition, as with [5], our simulation has included additive random noise that follows $\mathcal{N}(0, 0.01)$ to simulate imperfect reliability in actual PUF circuits.

ECP-TRN models were studied on 128-bit and 64-bit PUF variants. Random challenges were generated (in sets of 1 million) and applied to PUF simulation models as inputs.

Modeling experiments were conducted on 64-bit and 128-bit PUF variants using general ECP-TRN model characteristics as shown in Table 5.1. A set of 1 million random 128-bit and 64-bit challenges were generated, and given as input to the

Table 5.2 Experimental measurements of uniformity and uniqueness for selected PUFs

PUF	Challenge size (bits)	Uniformity (%)	Uniqueness (%)
APUF	64	51.04	50.19
	128	50.34	50.35
4-XOR APUF	64	50.81	50.68
	128	51.12	49.71
5-XOR APUF	64	50.00	49.98
	128	50.03	50.32
6-XOR APUF	64	52.31	50.48
	128	51.84	51.64
7-XOR APUF	64	49.23	53.21
	128	49.51	52.75

Table 5.3 Important hyperparameter values used

Challenge Size (bits)	No. of XORs	Weight tensor Rank	Training batch Size ($\times 10^3$)	Learning Rate ($\times 10^{-4}$)
64	4	5	10	8
	5	10	10	1
	6	10	80	80
	7	400	4.8	100
	8	1000	300	100
128	4	11	10	50
	5	250	300	50
	6	500	100	500
	7	1500	50	100

simulation models of PUF. Table 5.2 shows the uniformity and uniqueness metrics for selected PUF variants; we obtained similar uniformity and uniqueness metrics for other PUF variants. It should be noted that while the 6-XOR APUF's uniformity was a bit poorer than that of the other PUF variants we considered, it was still acceptable (52%) and did not influence modeling accuracy results. Parity vectors were calculated for each challenge and then used with the corresponding responses as input to the ECP-TRN for training and testing. A generated total set of CRPs for each PUF was divided into two parts: 80% of the CRPs were used as a training set and 20% as a validation set. Python 2.7 and TensorFlow [6] framework were used to implement the ECP-TRN modeling setup. All the experiments were run on an Ubuntu workstation with 64 GB main memory and a single core processor running at 3.3 GHz. The model is trained using *Backpropagation Algorithm* with the hyperparameters used which are listed in Table 5.3. For the benefit of readers, authors have made the source code available online[1].

5.4.2 Modeling Accuracy Results

ECP-TRN results report the minimum number of CRPs required to successfully model the PUF variant. CP-decomposed weight vectors are learned, with estimated rank parameters giving a low-rank approximation of the weight tensor. In the previous section, it was stated that the precise determination of tensor rank is difficult since tensor rank has been proven to be an NP-complete problem. The theoretical upper bounds of tensor rank were presented in Sect. 5.2.1.

The authors of the work [19] report that training TRN models are extremely sensitive to hyperparameter values. This can be attributed to the unconstrained CP-decomposition with backpropagation being one of the reasons of hyperparameter sensitivity [10]. Due to this, the hyperparameters required fine-tuning through

---

[1] https://github.com/Praneshss/Tensor-Regression-based-Modeling-Attack-on-PUFs.

5.4 Experimental Results

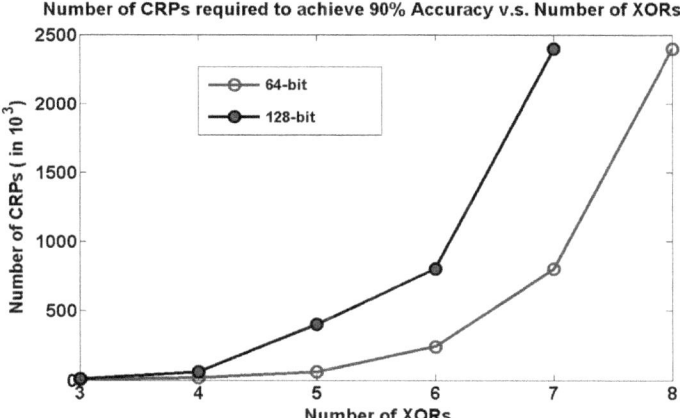

Fig. 5.3 Scaling of CRP training dataset size against number of XORs in the modeled XOR APUF, to achieve 90% modeling accuracy

multiple trials in each case. Table 5.3 lists the hyperparameters for modeling 64-bit and 128-bit XOR PUFs. *Batch size*, *learning rate*, and *tensor rank* are identified as the parameters that require careful tuning. A properly set λ parameter value for the L2 regularization will prevent overfitting of ECP-TRN. The ECP-TRN convergence depends largely on the tensor rank. In spite of the exponential dependence between tensor rank and dimensions of the tensor, the table shows that it does not exceed 1500, was set for the 128-bit 7-XOR APUF, and is considerably lower than the actual upper bound.

A comparison of the results obtained from modeling 64-bit and 128-bit XOR APUFs using the ECP-TRN technique, and other two previous works [2, 9], is presented in Table 5.4. Table includes required number of CRPs and prediction accuracy achieved for 64-bit and 128-bit XOR PUFs. Successful modeling was carried out until 8-XOR for 64 bit and 7-XOR for 128 bit. With superior computing infrastructure and more training data, the ECP-TRN attack can also be launched on higher x-values in x-XOR APUF. However, it is also considered that circuits with higher XOR-ings are very unreliable and severely impractical to use [4]. From the table, it is apparent that ECP-TRN outperforms previously used LR techniques regarding training time and data requirements, with consistently better or comparable results. The relationship between CRP training dataset size and the number of XORs in the modeled XOR APUF to achieve 90% modeling accuracy is shown in Fig. 5.3. The ECP-TRN model was tested to determine the robustness of the technique against noisy (erroneous) training data by introducing noise to the training dataset. The noise was imposed by randomly flipping some bits in the response data. Table 5.5 reports the accuracy results for noisy training dataset with 1%, 2%, and 5% error rates. Based on results of the table, it is clear that accuracy degrades gracefully with an increase in error rate in the noisy training data.

Table 5.4 Comparison of model accuracy results between ECP-TRN and LR for XOR APUF

Challenge Size (bits)	No. of XORs	Attack Method	Training CRP Count ($\times 10^3$)	Training Time	Best prediction Accuracy (%)
64	4	ECP-TRN	40	2 min	98.27
		[9]	42	16 sec	98.00
		[2]	12	3.42 min	99.00
	5	ECP-TRN	80	18 min	98.09
		[9]	260	2.46 min	98.00
		[2]	80	2.08 hrs	99.00
	6	ECP-TRN	320	40.20 min	97.39
		[9]	1400	30.34 min	98.00
		[2]	200	31.01 hrs	99.00
	7	ECP-TRN	560	6.2 hrs	97.78
		[9]	20000	2.43 hrs	98.00
		[2]	NA	NA	NA
	8	ECP-TRN	2700	15.5 hrs	97.42
		[9]	150000	6.31 hrs	98.00
		[2]	NA	NA	NA
128	4	ECP-TRN	60	5 min	98.15
		[9]	200	2.24 min	98.00
		[2]	24	2.53 hrs	99.00
	5	ECP-TRN	400	1.5 hrs	97.18
		[9]	2200	12.11 min	98.00
		[2]	500	16.36 hrs	99.00
	6	ECP-TRN	1200	6.1 hrs	97.76
		[9]	15000	4.45 hrs	98.00
		[2]	NA	NA	NA
	7	ECP-TRN	2800	18.2 hrs	96.65
		[9]	40000	66.53 hours	98.00
		[2]	NA	NA	NA

All the delay values for the arbiter chains in the experiments were selected from the same statistical distribution for a XOR APUF instance. ECP-TRN modeling attacks were repeated on 64-bit 5-XOR and 6-XOR APUFs, selecting random delay values from normal distributions with different mean (μ) and standard deviation (σ) values for each of the constituent arbiter chains to make sure that correlation effects are not a factor. Table 5.6 lists the different μ and σ values used. The same set of hyperparameters, mentioned in Table 5.3, were utilized for the 64-bit 5-XOR and 6-XOR APUFs and achieved modeling accuracy up to 97% at about the same computational effort. Therefore, it was concluded that ECP-TRN modeling attack experiments were free from delay correlation effects between arbiter chains.

5.5 ECP-TRN-Based Modeling Attack XOR APUF Variants

Table 5.5 Modeling accuracy results for XOR APUF on noisy training data

Challenge Size (bits)	No. of XORs	Prediction accuracy (%) at different error rates		
		(1% error)	(2% error)	(5% error)
64	4	96.30	93.7	91.40
	5	94.70	92.2	89.60
	6	94.30	91.60	82.50
	7	95.40	92.40	86.70
	8	85.00	79.00	71.00
128	4	96.60	92.50	88.10
	5	93.50	88.40	80.00
	6	96.30	94.50	89.80
	7	91.10	88.40	82.70

Table 5.6 Delay parameters of constituent APUFs (mean and standard deviation of normal distribution)

Distribution no.	Mean (μ)	Standard deviation (σ)
1	1.76	1.09
2	3.75	0.79
3	2.38	0.86
4	4.72	1.02
5	2.61	0.90
6	1.10	1.20

5.5 ECP-TRN-Based Modeling Attack XOR APUF Variants

To increase their resistance to machine-learning-based modeling attacks, many modifications have been proposed to the first proposed "classic" XOR APUF [3]. We have seen such different PUF constructions that uses XOR as a combination function in Sect. 2.3. Among the various XOR APUF variants, we will examine how ECP-TRN can be applied to LSPUF, PC-XOR APUF, and a special APUF variant referred to as mixed challenge XOR APUF.

All three approaches are based on variation of input challenge for each constituent APUF. In LSPUF [4], an applied challenge vector is transformed into multiple, non-identical challenge vectors, while in PC-XOR APUF [7], the challenges are generated using an independent pseudo-random number generator without any input applied challenge vector. A mixed challenge XOR APUF [9] uses a mixture of the master challenge and challenges derived from a PRNG for the constituent APUF. Next, we will discuss the ECP-TRN-based modeling attack on these three variants.

5.5.1 ECP-TRN-Based Modeling Attack on Lightweight Secure PUF (LSPUF)

It is easy to adapt the classic ECP-TRN model to perform modeling attack on LSPUF. When calculating the inner product with $\mathbf{w}_{rx}$ in Eq. (5.12) for the x-XOR APUF, the same parity vector Φ is used. However, when calculating the inner product with LSPUF, a custom parity vector must be generated from the transformed challenge for each APUF chain. The modified ECP-TRN equation can be expressed as

$$f(\mathcal{X}) = \sum_{r=1}^{R} \langle \Phi_1, \mathbf{w}_{r1} \rangle \cdots \langle \Phi_\mathbf{k}, \mathbf{w}_{rk} \rangle \cdots \langle \Phi_\mathbf{x}, \mathbf{w}_{rx} \rangle \cdot O_r + b \qquad (5.14)$$

where $\Phi_\mathbf{k}$ represents the transformed parity vector associated with kth APUF chain. To model 64-bit LSPUFs with Q constituent APUFs, Eq. (5.14) was used. The value of Q varied between 4 and 6 and the experimental setup was same as mentioned in Sect. 5.4. The modeling attack results presented in Table 5.7 are compared with previously reported results from [2, 9]. As can be seen from Table 5.7, the modified ECP-TRN achieves similar accuracy with less CRPs than previously reported works. The experiments were run on a basic workstation where 3 million CRPs were able to run maximum. Attempts to model higher order LSPUF ($x > 6$) were unsuccessful with the limitations of basic workstation. In accordance with previously reported modeling attempts for LSPUF [2], the LSPUF results suggest that it is more resistant to modeling attack than the classic XOR APUF.

Table 5.7 Comparison of modeling accuracy results between ECP-TRN and LR for 64-bit LSPUF

No. of XORs	Tensor Rank	Batch Size	Attack Method	Training CRP Count ($\times 10^3$)	Training Time	Best prediction Accuracy (%)
4	3	1200	ECP-TRN	32	3.35 min	97.67
			[9]	30	3.57 min	98.00
			[2]	12	1.28 hr	99.00
5	6	3500	ECP-TRN	160	10.30 min	97.21
			[9]	300	3.03 hr	98.00
			[2]	300	13.06 hr	99.00
6	15	10000	ECP-TRN	480	25.12 min	96.73
			[9]	1000	42.30 min	98.00
			[2]	–	–	–

5.5 ECP-TRN-Based Modeling Attack XOR APUF Variants

Table 5.8 Modeling accuracy results for PC-XOR APUF

No. of XORs	Tensor Rank	Batch Size	Training CRP Count ($\times 10^3$)	Training Time	Best prediction Accuracy (%)
4	1	2500	25	2.30 min	98.27
5	10	5000	480	48.20 min	98.09
6	–	–	3000	–	–

5.5.2 ECP-TRN-Based Modeling Attack on PC-XOR APUF

For the PC-XOR PUF, the modified ECP-TRN modeling with separate parity vectors for individual APUFs is applicable, just as it is for LSPUF modeling. Accordingly, this PUF can be attacked with modified ECP-TRN that is based on Eq. (5.14), with sub-challenges coming from software-implemented Linear Feedback Shift Registers (LFSRs), as used in [7].

The authors of [7] previously presented the modeling attack results for PC-XOR APUF with mention of the minimum number of CRPs that are requisite for the LR with resilient backpropagation (LR-RPROP) training algorithm to converge, but did not provide any details on the accuracy. For LR-RPROP to converge, the authors reported that a minimum of 8,500 CRPs are needed for 4-PC-XOR, 100,000 for 5-PC-XOR, and 16 million for 6-PC-XOR. The computational time required for the individual attack was not reported, but an aggregate of 100 days of computation time was mentioned for the attack.

Table 5.8 lists the results for ECP-TRN-based modeling attack on PC-XOR APUF. In accordance with the minimum CRPs mentioned by authors [7], the results appear consistent. We can see that attempts to model 6-PC-XOR with 3 million CRPs failed and the results also indicate that the modeling approach is more robust for challenges involving pseudo-random input compared to the classic XOR APUF (ref. Table 5.4).

5.5.3 ECP-TRN-Based Modeling Attack on Mixed Challenge XOR APUF

The mixed challenge XOR APUF combines a mixture of the applied master challenge and PRNG outputs to be applied to constitute APUFs and then XOR-ed the output response of constitute APUFs. ECP-TRN can be applied to such mixed challenge XOR APUFs (MCXOR APUF), based on Eq. (5.14), in order to examine the robustness to machine-learning-based modeling attacks. The ECP-TRN-based modeling attack has been investigated for 5-MCXOR APUF, whereas authors of [9] show results for 4-MCXOR APUF. This modeling attack experiment used a fixed batch size of 10,000 and tensor rank of 10. In Table 5.9, we report the results of

Table 5.9 Modeling accuracy results for 64-bit mixed challenge 5-XOR APUF

Fraction of pseudo-random Challenge	Training CRP Count in ($\times 10^3$)	Training Time	Best prediction Accuracy (%)
1/5	144	12.30 min	97.56
2/5	270	22.22 min	96.89
3/5	375	33.15 min	96.73
4/5	440	40.28 min	96.61

our modeling attack to achieve about 97% accuracy. This table shows the results for 64-bit 5-MCXOR APUF, with differing fractions of pseudo-random inputs (for example, a fractional value of 2/5 indicates a pseudo-random challenge was applied to two of the five constituent APUFs, etc.). From the results presented in the table, we observe that despite the fact that input challenge transformation improves ML attacks in general, the hardware overhead associated with the input transformation network and PRNG neutralizes the advantage of low hardware footprint.

In summary, we discussed the concept of designing and implementing a computationally and memory-efficient modeling attack method. However, have you ever thought that having an efficient ML model of a PUF (say the arbiter PUF) can be used in constructive applications, such as for authentication purposes. In our next chapter, we will explore the constructive aspects of modeling attacks. Keep reading!

References

1. Lim, D. (2004). *Extracting secret keys from integrated circuits*. Master's thesis, Massachusetts Institute of Technology, U.S.A.
2. Rührmair, U., Sehnke, F., Sölter, J., Dror, G., Devadas, S., & Schmidhuber, J. (2010). Modeling attacks on physical unclonable functions. In *Proceedings of the ACM Conference on Computer and Communications Security, Ser. CCS'10* (pp. 237–249).
3. Suh, G. E., & Devadas, S. (2007). Physical unclonable functions for device authentication and secret key generation. In *Proceedings of the Design Automation Conference, Ser. DAC'07* (pp. 9–14).
4. Majzoobi, M., Koushanfar, F., & Potkonjak, M. (2008). Lightweight secure PUFs. In *Proceedings of the IEEE/ACM International Conference on Computer-Aided Design, Ser. ICCAD'08* (670–673).
5. Sahoo, D. P., Mukhopadhyay, D., Chakraborty, R. S., & Nguyen, P. H. (2018). A multiplexer-based Arbiter PUF composition with enhanced reliability and security. *IEEE Transactions on Computers, 67*(3), 403–417.
6. Abadi, M., et al. (2016). TensorFlow: A system for large-scale machine learning. In *Proceedings of the USENIX Conference on Operating Systems Design and Implementation, Ser. OSDI'16* (pp. 265–283).
7. Yu, M.-D., Hiller, M., Delvaux, J., Sowell, R., Devadas, S., & Verbauwhede, I. (2016). A lockdown technique to prevent machine learning on PUFs for lightweight authentication. *IEEE Transactions on Multi-Scale Computing Systems, 2*(3), 146–159.

References

8. Brandi, G. (2018). *Decompose et impera: Tensor methods in high-dimensional data*. Ph.D. dissertation, LUISS Guido Carli.
9. Wisiol, N., Becker, G. T., Margraf, M., Soroceanu, T. A., Tobisch, J., & Zengin, B. (2020). Breaking the lightweight secure PUF: Understanding the relation of input transformations and machine learning resistance. *Smart Card Research and Advanced Applications* (pp. 40–54). Cham: Springer International Publishing.
10. Zhou, G., Cichocki, A., & Xie, S. (2014). Decomposition of big tensors with low multilinear rank. http://arxiv.org/abs/1412.1885.
11. He, L., Kong, X., Yu, P. S., Yang, X., Ragin, A. B., & Hao, Z. (2014). DuSK: A dual structure-preserving kernel for supervised tensor learning with applications to neuroimages. In M. J. Zaki, Z. Obradovic, P.-N. Tan, A. Banerjee, C. Kamath & S. Parthasarathy (Eds.), *SIAM International Conference on Data Mining* (pp. 127–135).
12. Khoromskij, B. N. (2018). *Tensor numerical methods in scientific computing* (Vol. 19). Walter de Gruyter GmbH & Co KG.
13. Kolda, T. G., & Bader, B. W. (2009). *Tensor decompositions and applications. SIAM review, 51*(3), 455–500.
14. Cao, X., Rabusseau, G., & Pineau, J. (2017). Tensor regression networks with various low-rank tensor approximations. *CoRR, abs/1712.09520*.
15. Håstad, J. (1989). Tensor rank is NP-complete. In *International Colloquium on Automata, Languages, and Programming* (pp. 451–460). Springer.
16. Hao, Z., He, L., Chen, B., & Yang, X. (2013). A linear support higher-order tensor machine for classification. *IEEE Transactions on Image Processing, 22*(7), 2911–2920.
17. Kossaifi, J., Khanna, A., Lipton, Z., Furlanello, T., & Anandkumar, A. (2017). Tensor contraction layers for parsimonious deep nets. In *IEEE Conference on Computer Vision and Pattern Recognition Workshops (CVPRW)* (pp. 1940–1946).
18. Kossaifi, J., Lipton, Z. C., Khanna, A., Furlanello, T., & Anandkumar, A. (2017). Tensor regression networks. *CoRR, abs/1707.08308*.
19. Santikellur, P., & Chakraborty, R. S. (2021). A computationally efficient tensor regression network based modeling attack on XOR Arbiter PUF and its variants. *IEEE Transactions on Computer-Aided Design of Integrated Circuits and Systems, 40*(6), 1197–1206.

Chapter 6
Combinational Logic-Based Implementation of PUF

6.1 Introduction

In the earlier chapters, we have noted that a standalone arbiter PUF is vulnerable to machine learning (ML) attacks; however, multiple instances of arbiter PUF can be combined to create more robust PUF variants. The security of these robust composite PUFs often depends on the basic building block, which is why it is important to analyze and develop new insights into the security of these building blocks.

In this regard, Boolean properties of PUF have been explored in recent years including correlation spectrum, juntas, and *Binarized Neural Network* (BNN)-based Boolean PUF representation. While the first two methods consider the PUF as blackbox Boolean mapping, the latter one derives Boolean function representation of the PUF. In this chapter, we examine the possibility of representing PUFs by Boolean functions using *Binarized Neural Networks* (BNNs) [3].

BNNs are a simplified form of a quantized neural network (QNN) whose weights and activation functions are represented in 1 bits. BNNs have much lower hardware overhead due to lightweight logic gates replacing bulky floating-point or fixed-point arithmetic circuitry of traditional neural networks. In addition, their computational latency is much lower. Thus, BNNs are promising candidates for silicon implementation of neural networks on those devices whose by design poses limited storage and computing power such as the Internet of Things (IoT). BNNs have been studied in a variety of ways, including a Boolean analysis and model verification [1, 8, 10, 11], improving training methods for learning complex datasets and efficient hardware implementations [5, 6].

The direct combinational gate-based circuit implementations of PUFs have not yet been investigated. BNNs can be used to achieve that. The focus of this chapter is particularly on achieving Boolean function representations of the arbiter PUF (APUF) using BNNs. This representation has several advantages:

1. The implementation and design of APUFs require careful placement of component switches and routing of the interconnections. In contrast, if a combinational logic gate representation of an APUF is available, it can be implemented easily on

any platform (FPGA, ASIC, microcontroller, etc.) without storing a CRP database explicitly or requiring new design techniques at the verifier end of a PUF-based authentication protocol. In particular, the implementation of BNNs on the verifier side is much simpler than machine learning models of APUF instance, especially when considering hardware implementation of PUF models.
2. This representation for an APUF can be regarded as a closed-form (albeit approximate) Boolean function representation, assisting their analysis and leading to potential reveal of the newer insights about the security of the instance. The framework is fully automated and achieves modeling accuracy up to 98% for BNN models, and a logic optimization process saves over 25% on hardware overhead.

It is beneficial if the direct combinational gate-based circuit implementations of PUF can be packaged in a framework where input is an arbiter PUF CRPs and output is a synthesizable Verilog description of combinational circuits corresponding to APUF instance. This chapter provides information on such a framework termed "APUF-BNN". The framework consists of two important phases: first, a modeling attack is launched on the APUF instance to generate a BNN model, and then the BNN model is optimized via a matrix-covering algorithm [13] to make the circuit implementation efficient. This framework has the advantage that it is fully automated, achieves higher accuracy, and saves hardware overhead through logic optimization.

The BNN model and the APUF instance behave differently due to the fact that they do not achieve 100% accuracy. Please note that this is similar to issues of imperfect reliability of the APUF circuitry, as no practical PUF circuit is 100% reliable. However, the reliability issue now lies with the verifier end at the prover end. There are some authentication protocols in which a small inaccuracy in the PUF model cannot impair the authentication process since they are robust in their nature. However, there are others that require a more accurate PUF model. These situations usually require PUF reliability enhancements generally achieved through Error Correction Code (ECC) such as *Helper Data Algorithm*, which make up the majority of commercially available solutions [14, 15] available today. Similarly, use of such solutions is also effective in overcoming the errors associated with the accuracy of the modeling. The following section discusses how the ML model of APUF can be used to achieve robust authentication with XOR APUF or other APUF compositions.

6.1.1 ML-Based PUF Models in PUF-Based Authentication Protocols

Several PUF-based authentication protocols utilize ML-based models in a constructive way, even though ML-based modeling attacks are seen as a potential threat. Although it is considered to be a potential threat, there are several proposals for PUF protocols [2, 4, 7] that utilize the model of a PUF instance in a constructive way.

It is typical for PUF protocols to propose using the compact ML model of the PUF for authentication purposes. This seems counterintuitive to the notions of PUF security at first glance, especially when the PUF variant being used is inherently robust against modeling attacks. In order to better understand this, consider the case of a practically robust x-XOR APUF, in which the final output is XOR of outputs of x different constituent APUFs, deployed on a device that must be authenticated. An XOR APUF has the XOR as combination function, which makes machine learning attacks on it difficult. In order to get the modeling the XOR APUF, each x-XOR APUF is modeled separately before deployment, so it is indirectly possible to construct the final XOR APUF model from constituent APUF models. Nevertheless, once XOR APUF is deployed in-field, the individual APUF output ports become inaccessible, which makes it difficult for an adversary to construct the model if they have physical access to the device that contains the PUF. **Based on similar assumptions, APUF-BNN considers PUF-based authentication protocols to be applicable and useful, especially when robust APUF compositions are employed**.

Prior to discussing the APUF-BNN framework, we briefly discuss BNNs in the following section, along with a matrix-covering algorithm for optimizing BNN implementations.

6.2 Binarized Neural Networks

The neural network parameters are typically represented using floating-point numbers. *Quantized Neural Networks* (QNNs) are neural networks that are characterized by low-precision weights and activations. QNNs are memory efficient and have little to no impact on the inference accuracy. *Binarized Neural Networks* (BNN) [3] are an extreme case of a quantized neural network, where weights and activations are represented by binary value instead of full-precision values. This facilitates fast computations since full-precision operation is converted to bitwise operations.

BNN constrains the network parameters to $\{-1, +1\}$. During inference using hardware implementations, the network parameters are represented using $\{0, 1\}$. Figure 6.1 demonstrates an example BNN. It has been reported that binarization can further assist in solving real-time image classification problems using *Convolutional Neural Networks* (CNN) [16]. In this chapter, we discuss feedforward fully connected BNNs, unless explicitly stated otherwise.

In contrast to post-training quantization methods, BNNs are trained from scratch. A well-known gradient-descent-based backpropagation algorithm is used to train the BNN. The backpropagation algorithm propagates through a neural network twice: once from the forward direction (*forward pass*) and the other from the backward direction (*backward pass*), *forward pass* and *backward pass*. In the forward pass, the output is computed by propagating forwardly using given inputs. The computed output value is compared with the actual target value and a *loss value* is derived. In the backward pass, the neural network weights are adjusted based on the loss value, so that the desired output is achieved. It is relatively easy to perform the

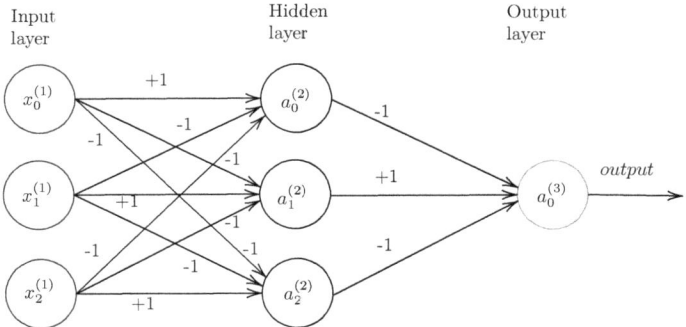

Fig. 6.1 Structure of a binarized neural network (BNN) with single hidden layer [3]

forward propagation and compute the loss value on a network with binary weights; however, performing gradient descent is not possible with binary weights. Hence, real-valued weights are yet used for gradient descent optimization. However, weights are binarized in the forward pass and loss value is computed from the binarized weights. This ensures the real-valued weights are not required during inference too, thereby making the neural network weights fully binary. Using the Sign() function, we can binarize a real-valued weight x_R that has the following correspondence:

$$\text{Sign}(x_R) = \begin{cases} 1 & \text{if } x_R \geq 0, \\ -1 & \text{otherwise} \end{cases} \quad (6.1)$$

However, the gradient of the Sign() function can be zero or undefined, which can be overcome by the heuristic approach termed *Straight-Through Estimator* (STE) [3]. This is alternate gradient propagation method disregards gradient calculation of the Sign() function during backward pass and solves it by assigning the inward gradient of Sign() to the outward gradient. It can be regarded as an identity function, which just passes the gradients. Although the STE seems somewhat arbitrary, theoretical reasons for it have been provided in multiple papers, e.g., in [7]. We suggest readers refer to [3] if they wish to know more about the BNN training algorithm.

The $\{-1, +1\}$ weight representation can be replaced by an alternative $\{0, +1\}$. In our terminology, *bipolar weights* are represented by their $\{-1, +1\}$ representation, whereas *encoded weights* are represented by their $\{0, 1\}$ representation. To illustrate, the key operation during inference is to compute the dot product of two vectors. Using the encoded representation, dot product becomes functionally equivalent to the summation of the XNOR operation outputs. Their correspondence is presented in Table 6.1. We now discuss the mathematical operations associated with BNN inference.

Let **x** be the output of layer l, and let the encoded weight matrix between two consecutive fully connected layers l and $l+1$ of BNN be W^l. Then, the output of the layer l is expressed as

6.3 Optimized Combinational Logic-Based BNN Implementation

Table 6.1 Relationship between bipolar multiplication and binary XNOR

Binary Weights		Encoded Weights		Multiplication Values	Bitwise XNOR
−1	−1	0	0	1	1
−1	1	0	1	−1	0
1	−1	1	0	−1	0
1	1	1	1	1	1

$$\mathbf{y} = g^{[l+1]}(W^l \odot \mathbf{x}) \tag{6.2}$$

where $W^l \odot \mathbf{x}$ represents XNOR matrix multiplication and $g^{[l+1]}$ is the activation function for layer $l + 1$. Please note that XNOR matrix multiplication is defined as

$$\begin{pmatrix} y_1 \\ y_2 \\ \vdots \\ y_m \end{pmatrix} = \begin{pmatrix} w_{11} & w_{12} & \cdots & w_{1n} \\ w_{21} & w_{21} & \cdots & w_{2n} \\ \vdots & \vdots & \ddots & \vdots \\ w_{m1} & w_{m2} & \cdots & w_{mn} \end{pmatrix} \odot \begin{pmatrix} x_1 \\ x_2 \\ \vdots \\ x_n \end{pmatrix} = \begin{pmatrix} (w_{11} \odot x_1) + \cdots + (w_{1n} \odot x_n) \\ (w_{21} \odot x_1) + \cdots + (w_{2n} \odot x_n) \\ \vdots \\ (w_{m1} \odot x_1) + \cdots + (w_{mn} \odot x_n) \end{pmatrix} \tag{6.3}$$

where m corresponds to the number of output neurons $\{v_1, v_2, \ldots, v_m\}$, and n corresponds to number of inputs $\{x_1, x_2, \ldots, x_n\}$. There can be two different BNN hardware implementation approaches corresponding to Eq. (6.3): *Parametric implementation* and *direct logic* implementation. In the parametric implementation, the activation outputs and weights are stored in memory for faster computation.

To make computation easier, the parametric implementation typically stores the activation outputs and weights into memory. Contrary to this, direct logic implementation is a pure combinational circuit where parameters are considered constants. On the other hand, the direct logic implementation is a pure combinational circuit implementation where parameters are hard-coded as constants, resulting in substantial hardware savings through logic optimization. These implementations also have higher performance results and impart lower circuit power dissipation. To implement BNN algorithms for APUF, APUF-BNN makes use of the second approach. Next, we discuss the implementation method in detail.

6.3 Optimized Combinational Logic-Based BNN Implementation

Instead of using memory elements for BNN parameter storage, our goal here is to find an optimized hardwired logic circuit for fully connected layers of BNN. Typically, at fully connected layers, there are more neurons, and the neuron inputs are shared

	1	2	3	4	5	6	7	8	9	10
1	0	0	1	1	0	0	1	1	0	0
2	0	1	1	1	0	0	1	0	1	0
3	0	1	1	0	0	1	1	0	1	0
4	0	0	1	0	0	0	1	1	1	0

Fig. 6.2 An encoded weight matrix W^l

among all the neurons in the hidden layer, which provides a strong argument to optimize based on logic sharing.

Based on the fact that BNN parameters are binary constants, we can optimize the network using the following algorithm. The bitwise XNOR operation $t_{ijk} = (w_{ij} \odot x_k)$ in Eq. (6.3), where $i \in \{1, \ldots, m\}$ and $j, k \in \{1, \ldots, n\}$, can be reduced to

$$t_{ijk} = \begin{cases} \neg x_k & \text{if } W^l_{ij} = 0, \\ x_k & \text{otherwise} \end{cases} \qquad (6.4)$$

Logic optimization is the process of finding the maximum region sharing the logic among rows in weight matrix W^l. A *matrix-covering* approach has been suggested in [13] as a way of solving this problem. The algorithm presented in [13] follows the row pairing principle and successfully achieves effective logic sharing among neurons in the fully connected layer of a BNN, resulting in a highly optimized hard-wired BNN circuit. As demonstrated in experimental results of [13], using this logic optimization method reduces the number of *Lookup Tables* (LUTs) and interconnect ("net") cost substantially for FPGA-based implementation of fully connected layers. In general, for a given weight matrix W^l of dimension $m \times n$, the fundamental principle is a common sub-pattern between two rows i_1 and i_2 appearing in columns $C \subseteq \{1, 2, \ldots, n\}$ of W^l weight matrix corresponding to a partial bit-counting circuit that is shared between two neurons v_i and v_j.

Let us understand the basic working principle of this method. Consider the weight matrix W^l of dimension $m \times n$. Now, a common sub-pattern between two rows i_1 and i_2 appearing in columns $C \subseteq \{1, 2, \ldots, n\}$ of W^l represents the partial bit-counting circuit shareable between the two corresponding neurons v_i and v_j. In order to better understand this, consider the weight matrix illustrated in Fig. 6.2. Here, columns represent input neurons $\{x_1, x_2, \cdots x_{10}\}$ and rows represent output neurons $\{v_1, v_2, v_3, v_4\}$, and partial summation is calculated by applying Eqs. (6.3) and (6.4) and includes these steps:

6.4 The APUF-BNN CAD Framework

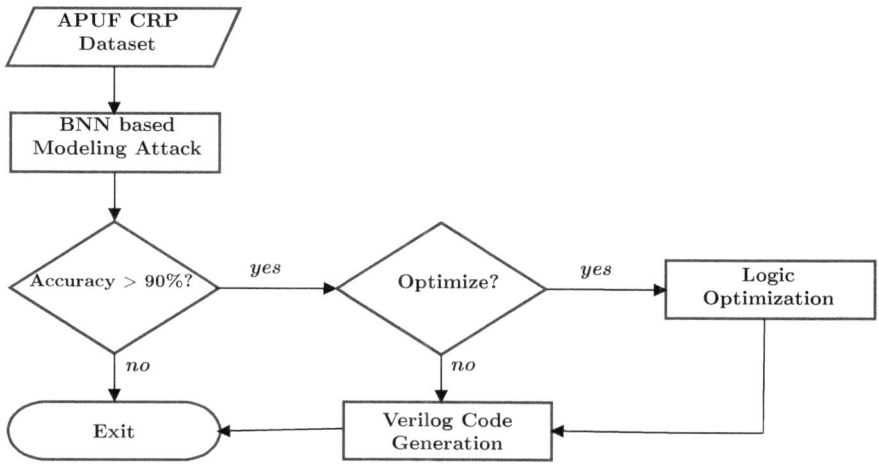

Fig. 6.3 The APUF-BNN framework-based design flow

- Neurons v_1, v_2, v_3, v_4 share bit pattern $\{\neg x_1, x_3, \neg x_5, x_7, \neg x_{10}\}$, which is equivalent to the partial summation $(\neg x_1 + x_3 + \neg x_5 + x_7 + \neg x_{10})$.
- Neurons v_2, v_3 share bit pattern $\{\neg x_2, \neg x_8, x_9\}$, which is equivalent to the partial summation $(\neg x_2 + \neg x_8 + x_9)$. Also, partial summation $(x_4 + \neg x_6)$ of neuron v_2 can be shared by neuron v_3 as $2 - (x_4 + \neg x_6)$, since these bit patterns complement each other. Thus, bit counting of $v3$ can be substituted by $2 - v2$.
- Neurons v_1, v_4 share bit pattern $\{\neg x_2, \neg x_6, x_8\}$, which amounts to the partial summation $(\neg x_2 + \neg x_6 + x_8)$. Also, partial summation $(x_4 + \neg x_9)$ of neuron v_1 can be shared by neuron v_4 as $2 - (x_4 + \neg x_9)$.

Matrix covering-based optimization heuristic achieves its objective by performing logic sharing steps repeatedly and identifying common sub-circuits shared by multiple neurons in the fully connected layer. In the next section, we will present the APUF-BNN CAD framework.

6.4 The APUF-BNN CAD Framework

The complete automated design flow followed by the APUF-BNN framework is shown in Fig. 6.3. There are three main steps. As a first step, a CRP dataset is used as an input to launch a BNN-based modeling attack on the given APUF instance. If the achieved accuracy of the BNN-based modeling is greater than 90%, the corresponding model parameters are supplied to the next step, i.e., logic optimization, otherwise the attempt is aborted. In case of an aborted attempt, a different dataset of CRPs, perhaps involving a higher number of CRPs, would be needed. Having successfully completed the step of achieving accurate BNN model, the logic optimization steps as described in the preceding section, it produces a list of shared logic neurons based on the BNN model parameters. Using this list, the final step of the framework generates synthesizable Verilog descriptions of a combinational circuit corresponding to the given APUF instance. Next, we describe each step of the framework.

6.4.1 BNN-Based Modeling Attack on APUF

The first step in the framework is to perform the modeling attack and create BNN model that corresponds to a given PUF instance. The *Linear Additive Delay Model* [1, 2], which has been discussed in Sect. 4.2, can be used to model the responses of an APUF instance. Relationship between response of an APUF instance and parity vector is described by Eq. (4.5). A BNN model is developed using supervised learning, in which the parity vector is used as an input and the response is used as a target label. As part of the training process of BNN, binarization functions and *Straight-Through estimator* (STE) algorithm are applied to restrict the weights to $-1, +1$. After achieving high accuracy (say, 90% accuracy), the binarization of APUF instances can be said to be successful. **Please note that bipolar encoding is used to represent parity vectors, which benefits the complete binarized representation of an APUF instance.** After a successful BNN model is built, the weight matrices are extracted and supplied to the next step of the framework, for logic optimization.

6.4.2 Matrix Covering-Based Logic Optimization and Combinational Verilog Code Generation

The second and third steps of the framework are logic optimization and Verilog code generation. With the trained BNNs from the first step, which corresponds to the given APUF instance, neuron weights are extracted. In fact, it is possible to directly skip step 2 and go directly to step 3, generating Verilog code for the circuit without optimization. The Verilog code can be generated by allocating a wire for each neuron and assigning its value to the sum of all inputs, with or without inversion. However, for an optimized representation of the circuit, the matrix-covering heuristic is applied to trained weights of each fully connected BNN layer of the APUF model which outputs a list of row of neurons with same input logic. Row pairs are assigned to intermediate wires, then their values are added in order to recover their original neuron outputs. Due to the fact that the circuit implementation is functionally identical to the mathematical BNN model, the Verilog implementation does not lose accuracy.

6.5 Experimental Results

6.5.1 Setup

We followed the same setup as described in Sect. 5.4.1. The study collected 100,000 CRPs from ten APUF instances in total, with the mean uniformity and mean uniqueness of each APUF being 50.03 and 49.86, respectively. Hence, all results (e.g., modeling accuracy, hardware footprint, etc.) are averaged across these ten APUF instances.

The APUF-BNN framework was implemented mainly using Python 2.7 and ML frameworks such as *Tensorflow* and *Keras* (version 2.1.5). APUF CRP datasets were split 80:20 between training and test sets, respectively. Implementation of matrix-covering-based logic optimization was done using *numpy* and *networkx* Python libraries and Bash shell scripting was used to automate the design flow. *Glorot initialization* [3] was used to initialize the BNN weights. *Adam optimizer* was used to update the weights and initial learning rate was set to 0.0001. The BNN model was trained for 100 epochs with a batch size of 800 and a decay rate of 0.95 for every 10 epochs as described in [3]. *Early stopping* was adopted to control the overfitting and over-training problems.

A basic workstation with 8 GB of main memory and an 8-core *Intel i7* processor was used to evaluate the framework. The Verilog code was implemented using *Vivado* FPGA software from Xilinx, and it was mounted onto a *Digilent Nexys-4* FPGA board. Over a serial communication link, APUF input challenges and corresponding output responses were communicated between the FPGA and the workstation.

6.5.2 Experimental Results

Five different BNNs were used for 64-bit and 128-bit APUF modeling, differing in the number of hidden layers and number of neurons in the hidden layer(s). These BNNs were named as (M_1, M_2, M_3, M_4, M_5). The accuracy of these five BNNs for modeling 64-bit and 128-bit APUF is shown in Fig. 6.4. It also shows the BNN architecture for modeling an n-bit APUF, with the number of neurons used at each layer denoted by "-". As an example, the BNN M_1 has three layers with ($n + 1$) neurons at an input layer, a hidden layer of 256 neurons, and an output layer with 1 neuron. At the input layer, ($n + 1$) neurons correspond to the bit length of the parity vector. Figure 6.1 also shows that with increasing parameter counts, modeling accuracy initially increases to some degree, but the change becomes marginal with increasing parameter counts. When parameter counts increase, corresponding circuit implementations require a larger hardware footprint. Therefore, it is important to find an architecture that offers a right balance between parameter count and test accuracy. In order to avoid an over-parameterized model, we chose three BNNs, each of which has only one hidden layer.

BNN Model	BNN Architecture
M_1	$(n+1)$-256-1
M_2	$(n+1)$-512-1
M_3	$(n+1)$-1024-1
M_4	$(n+1)$-1024-256-1
M_5	$(n+1)$-1024-512-1

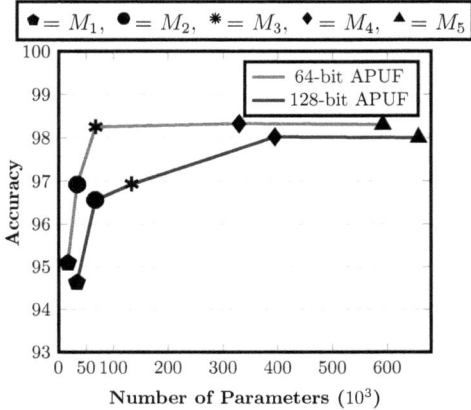

Fig. 6.4 Accuracy vs. number of parameters for 64-bit and 128-bit APUF. M_1 through M_5 correspond to different BNN models used to model an n-bit APUF, whose architectural details are given in the accompanying table

Table 6.2 Modeling accuracy and hardware footprint of BNN model architectures

PUF Size	BNN Model	Modeling Accuracy (%)	LUT count (unopt.)	LUT count (opt.)	LUT Savings (%)	Execution Time (mins.)
64 bit	M_1	95.39	11553	8748	24.27	9.06
	M_2	95.72	26262	19251	26.69	11.64
	M_3	98.25	52803	39445	25.29	18.82
128-bit	M_1	95.13	25483	16935	33.54	7.58
	M_2	96.56	50130	34981	30.21	9.67
	M_3	96.93	109666	70211	35.97	17.97

For each of the hardware implementations of the three BNN models (M_1, $M2$, and $M3$), Table 6.2 shows the results of the accuracy and footprint of the models. As can be seen in the table, BNN modeling test accuracy increases with the increase in the hidden layer neurons for both 64-bit and 128-bit APUFs, and 64-bit APUFs can achieve up to 98.25% accuracy.

The hardware footprints of the unoptimized and optimized BNNs were also reported in Table 6.2. The hardware footprints of a conventional, unoptimized BNN implementation were presented as a way to demonstrate the effectiveness of this logic optimization. As can be seen in the table, the logic optimization heuristic can significantly reduce hardware overhead ranging from 24.27% up to 35.97%.

A crucial criterion is the execution time of the APUF-BNN framework-enabled design flow. It consists of two major components that include training time for BNN-based APUF modeling, as well as the time for logic optimization and Verilog code generation. From the table, we see that for both 64-bit and 128-bit APUF, execution time increases with an increase in hidden layer neurons.

It is important to note that FPGAs using 64-bit APUF implementations use only 128 LUTs approximately [4]. It may be evident from Table 6.2 that the amount of hardware required for the circuit derived from this framework is more than what is needed by an ordinary APUF instance mapped onto a FPGA. Though there is a higher hardware footprint associated with this, alternative hardware representation as compared to the APUF-BNN framework we developed, the value of this alternative hardware representation should be evident from Sect. 6.1.

In summary, an alternative efficient hardware representation of arbiter PUF can be achieved using BNN followed by a logic optimization method. Using such a representation, further research can be undertaken to assess the security of APUF-based compositions.

References

1. Choi, A., Shi, W., Shih, A., & Darwiche, A. (2019). Compiling neural networks into tractable boolean circuits. In *AAAI Spring Symposium on Verification of Neural Networks (VNN)*.
2. Gao, Y., Ma, H., Al-Sarawi, S. F., Abbott, D., & Ranasinghe, D. C. (2018). PUF-FSM: A controlled strong PUF. *IEEE Transactions on Computer-Aided Design of Integrated Circuits and Systems, 37*(5), 1104–1108.
3. Hubara, I., Courbariaux, et al. (2016). Binarized neural networks. In *Advances in Neural Information Processing Systems* (pp. 4107–4115).
4. Majzoobi, M. (2012). Slender PUF protocol: A lightweight, robust, and secure authentication by substring matching. In *IEEE Symposium on Security and Privacy Workshops* (pp. 33–44).
5. Darabi, S., Belbahri, M., Courbariaux, M., & Nia, V. P. (2018). Bnn+: Improved binary network training. http://arxiv.org/abs/1812.11800.
6. Ghasemzadeh, M., Samragh, M., & Koushanfar, F. (2018). ReBNet: Residual binarized neural network. In *IEEE 26th Annual International Symposium on Field-Programmable Custom Computing Machines (FCCM)* (pp. 57–64).
7. Yin, P., Lyu, J., Zhang, S., Osher, S., Qi, Y., & Xin, J. (2019). Understanding straight-through estimator in training activation quantized neural nets. http://arxiv.org/abs/1903.05662.
8. Riazi, M. S., Samragh, M., Chen, H., Laine, K., Lauter, K., & Koushanfar, F. (2019). XONN: XNOR-based oblivious deep neural network inference. In *28th USENIX Security Symposium* (pp. 1501–1518).
9. Yu, M.-D., Hiller, M., Delvaux, J., Sowell, R., Devadas, S., & Verbauwhede, I. (2016). A lockdown technique to prevent machine learning on PUFs for lightweight authentication. *IEEE Transactions on Multi-Scale Computing Systems, 2*(3), 146–159.
10. Shih, A., Darwiche, A., & Choi, A. (2019). Verifying binarized neural networks by angluin-style learning. In *International Conference on Theory and Applications of Satisfiability Testing* (pp. 354–370). Springer.
11. Shih, A., Darwiche, A., & Choi, A. (2019). Verifying binarized neural networks by local automaton learning. In *AAAI Spring Symposium on Verification of Neural Networks (VNN)*.
12. Chatterjee, U., Chakraborty, R. S., & Mukhopadhyay, D. (2017). A PUF-based secure communication protocol for IoT. *ACM Transactions on Embedded Computing Systems, 16*(3), 1–25.
13. Chi, C.-C., & Jiang, J.-H. R. (2018). Logic synthesis of binarized neural networks for efficient circuit implementation. In *IEEE/ACM International Conference on Computer-Aided Design (ICCAD)* (pp. 1–7).

14. INTRINSIC, ID. (2019). QuiddiKey: Unclonable identities for the IoT. Accessed October 2020, from https://www.intrinsic-id.com/products/quiddikey/.
15. Delvaux, J., et al. (2015). Helper data algorithms for PUF-based key generation: Overview and analysis. *IEEE Transactions on Computer-Aided Design of Integrated Circuits and Systems, 34*(6), 889–902.
16. Rastegari, M., Ordonez, V., Redmon, J., & Farhadi, A. (2016). XNOR-Net: ImageNet classification using binary convolutional neural networks. In *European Conference on Computer Vision* (pp. 525–542). Springer.

Chapter 7
Conclusion

We have focused on applications of machine learning in the domain of hardware security, with a particular cover on modeling attacks against PUFs in this book. In the first chapter, we have introduced briefly popular hardware-based attacks and countermeasures. Also, we have laid out how machine learning is different from traditional algorithms. In the later sections of the chapter, we discussed the various applications of ML on hardware security, including the attacks and countermeasures.

The second chapter discusses PUFs in depth. We have discussed different properties and quality metrics related to PUF in the early parts of the chapter. In the later parts of the chapter, we discussed the workings of arbiter PUF and its compositions in more detail. The second chapter not only explores the basic ideas of PUFs and their compositions but also introduces the state-of-the-art ones such as interpose PUFs and tribes PUFs.

Chapter 3 discusses the basics of machine learning essential to understanding modeling attacks on PUFs. Following up on the introduction to machine learning in the first chapter, we introduced the different forms of machine learning. The later parts of chapter cover different well-known unsupervised and supervised machine learning algorithms.

Following an overview of the prerequisites and basic concepts in Chaps. 1–4 offers a detailed and comprehensive discussion of modeling attacks. First, we demonstrated the importance of features for modeling attacks and show the derivation of features for modeling arbiter PUF. Next, we introduced different types of modeling attacks such as using an exact mathematical model or using black-box modeling, and discussed their pros and cons. Finally, we examined extensively deep feedforward neural networks (DFNN)-based and Logistic Regression (LR)-based modeling attacks on different APUF compositions.

In Chap. 5, we discussed an improved modeling attack called tensor regression-based modeling on XOR arbiter PUF. The chapter starts off by reviewing different tensor concepts and their decomposition techniques. Following that, the chapter explains how CP-decomposition and tensor regression concepts are applied to XOR APUF. Additionally, an extension of the application of tensor regression model to XOR APUF variants is also discussed. Toward the end of the chapter, we talked about the results and compare them with previous known techniques.

In Chap. 6, we focused on an interesting and constructive side of modeling attacks. We showed how it is possible to achieve combinatorial representation using an ML model of arbiter PUF. The chapter covered the basics of BNN, matrix-covering optimization, and the APUF-BNN framework, which takes input in the form of CRPs of the arbiter PUF and returns a combinatorial representation.

From the basics of PUF to the most recent modeling attacks in the field, this book acts as a valuable reference for students, academic researchers, and industry practitioners alike. We believe it will provide readers with comprehensive information on modeling attacks on PUFs. We also hope that the contents of the book would pave the way for future advances of this topic.

GPSR Compliance
The European Union's (EU) General Product Safety Regulation (GPSR) is a set of rules that requires consumer products to be safe and our obligations to ensure this.

If you have any concerns about our products, you can contact us on

ProductSafety@springernature.com

In case Publisher is established outside the EU, the EU authorized representative is:

Springer Nature Customer Service Center GmbH
Europaplatz 3
69115 Heidelberg, Germany

www.ingramcontent.com/pod-product-compliance
Ingram Content Group UK Ltd.
Pitfield, Milton Keynes, MK11 3LW, UK
UKHW021251180426
11946UKWH00003B/71